T0215548

Beginning DAX with Power BI

The SQL Pro's Guide to Better Business Intelligence

Philip Seamark

Apress®

Beginning DAX with Power BI

Philip Seamark
UPPER HUTT, New Zealand

ISBN-13 (pbk): 978-1-4842-3476-1
https://doi.org/10.1007/978-1-4842-3477-8

ISBN-13 (electronic): 978-1-4842-3477-8

Library of Congress Control Number: 2018937896

Copyright © 2018 by Philip Seamark

This work is subject to copyright. All rights are reserved by the Publisher, whether the whole or part of the material is concerned, specifically the rights of translation, reprinting, reuse of illustrations, recitation, broadcasting, reproduction on microfilms or in any other physical way, and transmission or information storage and retrieval, electronic adaptation, computer software, or by similar or dissimilar methodology now known or hereafter developed.

Trademarked names, logos, and images may appear in this book. Rather than use a trademark symbol with every occurrence of a trademarked name, logo, or image we use the names, logos, and images only in an editorial fashion and to the benefit of the trademark owner, with no intention of infringement of the trademark.

The use in this publication of trade names, trademarks, service marks, and similar terms, even if they are not identified as such, is not to be taken as an expression of opinion as to whether or not they are subject to proprietary rights.

While the advice and information in this book are believed to be true and accurate at the date of publication, neither the authors nor the editors nor the publisher can accept any legal responsibility for any errors or omissions that may be made. The publisher makes no warranty, express or implied, with respect to the material contained herein.

Managing Director, Apress Media LLC: Welmoed Spahr
Acquisitions Editor: Joan Murray
Development Editor: Laura Berendson
Coordinating Editor: Jill Balzano

Cover designed by eStudioCalamar

Cover image designed by Freepik (www.freepik.com)

Distributed to the book trade worldwide by Springer Science+Business Media New York, 233 Spring Street, 6th Floor, New York, NY 10013. Phone 1-800-SPRINGER, fax (201) 348-4505, e-mail orders-ny@springer-sbm.com, or visit www.springeronline.com. Apress Media, LLC is a California LLC and the sole member (owner) is Springer Science + Business Media Finance Inc (SSBM Finance Inc). SSBM Finance Inc is a **Delaware** corporation.

For information on translations, please e-mail rights@apress.com, or visit http://www.apress.com/rights-permissions.

Apress titles may be purchased in bulk for academic, corporate, or promotional use. eBook versions and licenses are also available for most titles. For more information, reference our Print and eBook Bulk Sales web page at http://www.apress.com/bulk-sales.

Any source code or other supplementary material referenced by the author in this book is available to readers on GitHub via the book's product page, located at www.apress.com/9781484234761. For more detailed information, please visit http://www.apress.com/source-code.

Printed on acid-free paper

To Grace, Hazel, Emily
. . . and of course Rebecca.

Table of Contents

About the Author

Philip Seamark is an experienced Data Warehouse and Business Intelligence consultant with a deep understanding of the Microsoft stack and extensive knowledge of Data Warehouse methodologies and enterprise data modeling. He is recognized for his analytical, conceptual, and problem-solving abilities and has more than 25 years of commercial experience delivering business applications across a broad range of technologies. His expertise runs the gamut from project management, dimensional modeling, performance tuning, ETL design, development and optimization, and report and dashboard design to installation and administration.

In 2017 he received a Microsoft Data Platform MVP award for his contributions to the Microsoft Power BI community site, as well as for speaking at many data, analytic, and reporting events around the world. Philip is also the founder and organizer of the Wellington Power BI User Group.

About the Technical Reviewer

Jeffrey Wang stumbled upon Power BI technologies by accident in 2002 in the suburbs of Philadelphia, fell in love with the field, and has stayed in the BI industry ever since. In 2004, he moved his family across the continent to join the Microsoft Analysis Services engine team as a software engineer just in time to catch the tail end of SQL Server 2005 Analysis Services RTM development. Jeffrey started off with performance improvements in the storage engine. After he found that users desperately needed faster MDX (Multidimensional) calculations, he switched to the formula engine and participated in all the improvements to MDX afterward. He was one of the inventors of the DAX programming language in 2009 and has been driving the progress of the DAX language since then. Currently, Jeffrey is a principal engineering manager in charge of the DAX development team and is leading the next phase of BI programming and its modeling evolution.

Foreword

Power BI has changed the definition of business intelligence (BI) tools. Taking tools that were historically reserved for data scientists and making them easy to use and economically available to business analysts has enabled an entire culture of self-service BI. This book is doing the same for data access programming. It breaks complex concepts into simple, easy-to-understand steps and frames them in a business context that makes it easy for you to start learning the DAX language and helps you solve real-world business challenges on a daily basis.

—Charles "Chuck" Sterling (of Microsoft).

Acknowledgments

Thanks to Marco Russo, Alberto Ferrari, and Chris Webb for providing many years of high-quality material in the world of business intelligence.

CHAPTER 1

Introduction to DAX

The aim of this book is to help you learn how you can use the DAX language to improve your data modelling capability using tools such as Microsoft Power BI, Excel Power Pivot, and SSAS Tabular. This book will be particularly useful if you already have a good knowledge of T-SQL, although this is not essential.

Throughout the book, I present and solve a variety of scenarios using DAX and provide equivalent T-SQL statements primarily as a comparative reference to help clarify each solution. My personal background is as someone who has spent many years building solutions using T-SQL, and I would like to share the tips and tricks I have acquired on my journey learning DAX with those who have a similar background. It's not crucial for you to be familiar with T-SQL to get the best out of this book because the examples will still be useful to someone who isn't. I find it can be helpful to sometimes describe an answer multiple ways to help provide a better understanding of the solution.

In this book, I use Power BI Desktop as my primary DAX engine and most samples use data from the WideWorldImportersDW database, which is freely available for download from Microsoft's website. This database can be restored to an instance of Microsoft SQL Server 2016 or later. I am using the Developer edition of SQL Server 2016.

I recommend you download and install the latest version of Power BI Desktop to your local Windows PC. The download is available from `powerbi.microsoft.com/desktop`, or you can find it via a quick internet search. The software is free to install and allows you to load data and start building DAX-based data models in a matter of minutes.

The WideWorldImportersDW database is clean, well-organized, and an ideal starting point from which to learn to data model using DAX.

The aim of this first chapter is to cover high-level fundamentals of DAX without drilling into too much detail. Later chapters explore the same fundamentals in much more depth.

© Philip Seamark 2018
P. Seamark, *Beginning DAX with Power BI*, https://doi.org/10.1007/978-1-4842-3477-8_1

What Is DAX?

Data Analysis Expressions (DAX) is both a query and functional language. It made its first appearance back in 2009 as part of an add-in to Microsoft Excel 2010. The primary objective of DAX is to help organize, analyze, understand, and enhance data for analytics and reporting.

DAX is not a full-blown programing language and does not provide some of the flow-control or state-persistence mechanisms you might expect from other programming languages. It has been designed to enhance data modeling, reporting, and analytics. DAX is constantly evolving with new functions being added on a regular basis.

DAX is described as a functional language, which means calculations primarily use functions to generate results. A wide variety of functions are provided to help with arithmetic, string manipulation, date and time handling, and more. Functions can be nested but *you cannot create your own*. Functions are classified into the following categories:

- DateTime

- Filter

- Info

- Logical

- Mathtrig

- ParentChild

- Statistical

- Text

There are over 200 functions in DAX. Every calculation you write will use one or more of these. Each function produces an output with some returning a single value and others returning a table. Functions use parameters as input. Functions can be nested so the output of one function can be used as input to another function.

Unlike T-SQL, there is no concept of INSERT, UPDATE, or DELETE for manipulating data in a data model. Once a physical table exists in a Power BI, SSAS Tabular, or Excel PowerPivot data model, DAX cannot add, change, or remove data from that table. Data can only be filtered or queried using DAX functions.

What Is a Data Model?

A *data model* is a collection of data, calculations, and formatting rules that combine to create an object that can be used to explore, query, and better understand an existing dataset. This can include data from many sources.

Power BI Desktop, SSAS Tabular, and PowerPivot for Excel can import data from a wide variety of data sources including databases and flat files or directly from many source systems. Once imported, calculations can be added to the model to help explore and make sense of the data.

Data is organized and stored into tables. Tables are two dimensional and share many characteristics with databases tables. Tables have columns and rows, and relationships can be defined between tables to assist calculations that use data from multiple tables. Calculations can be as simple as providing a row count over a table or providing a sum of values in a column. Well-considered calculations should enhance your data model and support the process of building reports and performing analytical tasks known as *measures*.

It's the combination of data and measures that become your data model.

The Power BI Desktop user interface consists of three core components. First, the *Report View* provides a canvas that lets you create a visual layer of your data model using charts and other visuals. Report View also lets you control the layout by dragging, dropping, and resizing elements on the report canvas. It's the canvas that is presented to the end user when they access the report.

The second component is the *Data View*, which provides the ability to see raw data for each table in the model. Data View can show data for one table at a time and is controlled by clicking the name of a table from the list of tables in the right-hand panel. Columns can be sorted in this view, but sorting here has no impact on any sorting by visuals on the report canvas. Columns can be renamed, formatted, deleted, hidden, or have their datatype defined using the Report View. A hidden column will always appear in the Report View but not in any field list in the report.

It's possible to add or change calculations from both Report and Data View.

The last component of the Power BI Desktop user interface is the *Relationship View*. This section shows every table in the data model and allows you to add, change, or remove relationships between tables.

Components of a DAX Data Model

The DAX data modeling engine is made up of six key components.

Data

The first step of building a data model is importing data. A wide variety of data sources are available, and once they are imported, they will be stored in two-dimensional tables. Sources that are not two dimensional can be used, but these will need to be converted to a two-dimensional format before or during import. The query editor provides a rich array of functions that help with this type of transformation.

Tables

Tables are objects used to store and organize data. Tables consist of columns that are made up of source data or results of DAX calculations.

Columns

Each table can have one or more columns. The underlying data engine stores data from the same column in its own separate index. Unlike T-SQL, DAX stores data in columns rather than in rows. Once data has been loaded to a column, it is considered static and cannot be changed. Columns can also be known as *fields*.

Relationships

Two tables can be connected via a relationship defined in the model. A single column from each table is used to define the relationship. Only *one-to-many* and *one-to-one* relationships are supported. Many-to-many relationships cannot be created. In DAX, the most common use of relationships is to provide filtering rather than to mimic normalization of data optimized for OLTP operations.

Measures

A *measure* is a DAX calculation that returns a single value that can be used in visuals in reports or as part of calculations in other measures. A measure can be as simple as a row count of a table or sum over a column. Measures react and respond to user interaction and recalculate as a report is being used. Measures can return new values based on updates to the selection of filters and slicers.

Hierarchies

Hierarchies are groupings of two or more columns into levels that can be drilled up/down through by interactive visuals and charts. A common hierarchy might be over date data that creates a three-level hierarchy over year, month, and day. Other common hierarchies might use geographical data (country, city, suburb), or structures that reflect organizational groupings in HR or Product data.

Your First DAX Calculation

It's possible to import data and have no need to write in DAX. If data is clean and simple and report requirements are basic, you can create a model that needs no user-created calculations. Numeric fields dragged to the report canvas will produce a number.

If you then drag a non-numeric field to the same visual, it automatically assumes you would like to group your numeric field by the distinct values found in your nonnumeric field. The default aggregation over your numeric field will be SUM. This can be changed to another aggregation type using the properties of your visual. Other aggregation types include AVERAGE, COUNT, MAX, MIN and so on.

In this approach, the report creates a DAX-calculated measure on your behalf. These are known as *implicit measures*. Dragging the 'Fact Sale'[Quantity] field to the canvas automatically generates the following DAX statement for you:

```
CALCULATE(SUM('Fact Sale'[Quantity]))
```

This calculation recomputes every time a slicer or filter is changed and should show values relevant for any filter settings in your report.

Most real-world scenarios require at least some basic enhancements to raw data, and this is where adding DAX calculations can improve your model. When you specifically create calculated measures, these are known as *explicit measures*.

Some of the most common enhancements are to provide the ability to show a count of the number of records in a table or to sum values in a column.

Other enhancements might be to create a new column using values from other columns in the same row, or from elsewhere in the model. A simple example is a column that multiplies values from columns such as Price and Qty together to produce a total. A more complicated example might use data from other tables in a calculation to provide a value that has meaning to that row and table.

Once basic count or sum calculations have been added, more sophisticated calculations that provide cumulative totals, period comparisons, or ranking can be added.

These are the three types of calculations in DAX:

- Calculated columns

- Calculated measures

- Calculated tables

We explore each of these calculations in more detail later in this book and I include hints on how and when you might choose one type over another.

Note   *Calculated tables* (tables that are the result of a DAX calculation) can only be created in the DAX engine used by Power BI Desktop and SSAS Tabular.

Your First Calculation

The first example I cover creates a simple calculated measure using data from the WideWorldImportersDW database (Figure 1-1). The dataset has a table called 'Fact Sale' that has a column called [Total Including Tax]. The calculation produces a value that represents a sum using values stored in this column.

Quantity	Unit Price	Tax Rate	Total Excluding Tax	Tax Amount	Profit	Total Including Tax	To
1	13	15	13	1.95	8.5	14.95	
1	13	15	13	1.95	8.5	14.95	
1	13	15	13	1.95	8.5	14.95	
1	13	15	13	1.95	8.5	14.95	
1	13	15	13	1.95	8.5	14.95	
1	13	15	13	1.95	8.5	14.95	
1	13	15	13	1.95	8.5	14.95	
1	13	15	13	1.95	8.5	14.95	
1	13	15	13	1.95	8.5	14.95	
1	13	15	13	1.95	8.5	14.95	
1	13	15	13	1.95	8.5	14.95	
1	13	15	13	1.95	8.5	14.95	
1	13	15	13	1.95	8.5	14.95	

Figure 1-1. *A sample of data from the 'Fact Sale' table*

When viewing this table in Data View, we see the unsummarized value for each row in the [Total Including Tax] column. A calculated measure is required to show a single value that represents a sum of every row in this column.

In Power BI, you can create a calculated measure using the ribbon, or by right-clicking the table name in the Report View or Data View. This presents an area below the ribbon where you can type the DAX code for your calculation. The text for this calculated measure should be

```
Sum of Total including Tax = SUM('Fact Sales'[Total Including Tax])
```

This should look as it does in Figure 1-2.

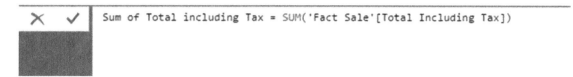

Figure 1-2. *DAX for the first calculated measure. This calculation uses the SUM function to return a single value anywhere the calculated measure is used in the report. When dragged and dropped to the report canvas using a visual with no other fields or filters, the value should show as $198,043,493.45.*

The structure of the formula can be broken down as follows: starting from the left, the first part of the text sets the name of the calculated measure. In this case, the name is determined by all text to the left of the = operator. The name of this calculated measure is [Sum of Total including Tax]. Names of calculated measures should be unique across the model including column names.

This name is how you will see the measure appear in the field list as well as how it may show in some visuals and charts.

Note   Spaces between words are recommended when creating names for calculated measures and columns. Avoid naming conventions that use the underscore character or remove spaces altogether. Use natural language as much as possible. Use names that are brief and descriptive. This will be especially helpful for Power BI features such as Q&A.

The = sign separates the calculation name from the calculation itself. A calculated measure can only return a single value and never a list or table of values. In more advanced scenarios, steps involving groups of values can be used, but the result must be a single value.

All text after the = sign is the DAX code for the calculated measure. This calculation uses the SUM function and a single parameter, which is a reference to a column. The single number value that is returned by the SUM function represents values from every row from the [Total Including Tax] column added together. The datatype for the column passed to the SUM function needs to be numeric and cannot be either the Text or DateTime datatypes.

The notation for the column reference is *fully qualified*, meaning it contains both the name of the table and name of the column. The table name is encapsulated inside single quotations (' '). This is optional when your table name doesn't contain spaces. The column name is encapsulated inside square brackets ([ ]).

Calculated measures belong to a single table but you can move them to a new home table using the Home Table option on the Modeling tab. Calculated measures produce the same result regardless of which home table they reside on.

Note   When making references to other calculated measures in calculations, never prefix them with the table name. However, you should always include the name of a table when referencing a column.

IntelliSense

IntelliSense is a form of predictive text for programmers. Most modern programming development environments offer some form of IntelliSense to help guide you as you write your code. If you haven't encountered this before, it is an incredibly useful way to help avoid syntax errors and keep calculations well-formed.

DAX is no different, and as you start typing your calculation, you should notice suggestions appearing as to what you might type next. Tooltips provide short descriptions about functions along with details on what parameters might be required.

IntelliSense also helps you ensure you have symmetry with your brackets, although this can sometimes be confusing if you are not aware it is happening.

IntelliSense suggestions can take the form of relevant functions, or tables or columns that can be used by the current function. IntelliSense is smart enough to only offer suggestions valid for the current parameter. It does not offer tables to a parameter that only accepts columns.

In the case of our formula, when we type in the first bracket of the SUM function, IntelliSense offers suggestions of columns that can be used. It does not offer tables or calculated measures as options because the SUM function is only designed to work with columns.

Formatting

As with T-SQL and pretty much any programming language, making practical use of line spacing, carriage returns, and tabs greatly improves readability and, more importantly, understanding of the code used in the calculation. Although it's possible to construct a working calculation using complex code on just a single line, it is difficult to maintain. Single-line calculations also lead to issues playing the *bracket game*.

Note   The *bracket game* is where you try to correctly pair open/close brackets in your formula to produce the correct result. Failure to pair properly means you lose the game and your formula doesn't work.

A good tip is to extend the viewable area where you edit your calculation before you start by repeating the Shift-Enter key combination multiple times or by clicking the down arrow on the right-hand side.

Comments

Comments can also be added to any DAX calculation using any of the techniques in Table 1-1.

Table 1-1. *How to Add Comments*

Comment Characters	Effect
//	Text to the right is ignored by DAX until the next carriage return.
--	Text to the right is ignored by DAX until the next carriage return.
/* */	Text between the two stars is ignored by DAX and comments can span multiple lines.

Examples of ways you can add comments are shown in Figure 1-3 as is an example of how spacing DAX over multiple lines helps increase readability.

```
Measure = DATEDIFF(
                -- This is the first argument
                TODAY() ,
                // This is the second argument
                DATE(2017,12,24),
                /* This is the final argument */
                DAY
                )
```

Figure 1-3. *Commented text styles and usage of line breaks*

By formatting code and adding comments you help to make the logic and intent of the function easier to understand and interpret for anyone looking at the calculation later.

A nice alternative to the formula bar in Power BI Desktop and SSAS Tabular, is DAX Studio, which is a free product full of features designed to help you develop and debug your calculations. I provide a more detailed look at DAX Studio in Chapter 8.

Your Second DAX Calculation

In this second example, I create a calculated column as opposed to a calculated measure. A Chapter 9 provides more detailed advice on when you should consider using a calculated column instead of a calculated measure, but in short, a calculated column adds a column in an existing table in which the values are generated using a DAX formula.

A simple calculation might use values from other columns in the same row. This example uses two columns from the 'Fact Sale' table to perform a simple division to return a value that represents an average unit price.

To add this calculation to your data model, select the 'Fact Sale' table in the Fields menu so it is highlighted. Then use the New Column button on the Modeling tab and enter the text shown in Figure 1-4.

```
Average Item Price = DIVIDE(
                          'Fact Sale'[Total Including Tax] ,
                          'Fact Sale'[Quantity]
                          )
```

Figure 1-4. A calculated column for [Average Item Price]

Note   This formula works without the table name preceding each column name; however, it is highly recommended that whenever you reference a column, you always include the table name. This makes it much easier to differentiate between columns and measures when you are debugging longer DAX queries.

The code in Figure 1-4 adds a new column to the 'Fact Sale' table called [Average Item Price]. The value in each cell of the new column (see Figure 1-5) is the output of this calculation when it is executed once for every row in the table.

Quantity	Total Including Tax	Average Item Price
90	1552.5	$17.25
90	3001.5	$33.35
90	3829.5	$42.55
90	10246.5	$113.85
90	3881.25	$43.13
90	3881.25	$43.13
90	2277	$25.30
90	3001.5	$33.35
90	4347	$48.30
90	3415.5	$37.95
90	2277	$25.30
90	1552.5	$17.25

Figure 1-5. *Sample of data for new calculated column*

Your Third DAX Calculation

This last example creates a calculated table. This option is only available in Power BI Desktop and SSAS Tabular, not in PowerPivot for Excel 2016 (or earlier).

You can create calculated tables using any DAX function that returns a table or by simply referencing another table in the model. The simplest syntax allows you to create a clone of another DAX table. The example shown in Figure 1-6 creates a new calculated table called 'Dates2', which is a full copy of the 'Dates' table. Modifications made to the 'Dates' table, such as adding or modifying columns, automatically flow through to the 'Dates2' table.

Figure 1-6. *Creating a calculated table*

Filters, measures, columns, and relationships can be added to the 'Dates2' table without effecting 'Dates' table. The beauty of this is that because the base table ('Dates') is reading from a physical data source, any changes to data in the 'Dates' table are reflected in the 'Dates2' table.

To extend the example so the new table only shows some rows from the original table, you can use the FILTER function as shown in Figure 1-7.

```
Dates2 = FILTER(
                Dates,
                'Dates'[Calendar Year]=2016
                )
```

Figure 1-7. *Calculated table created using FILTER*

The 'Dates2' calculated table still has the same number of columns as 'Dates', but it only has rows that match the filter expression. This results in the 'Dates2'calculated table having 365 rows that represent a row per day for the calendar year of 2016.

It is common to use calculated tables to create summary tables that can be used as faster alternatives for calculated measures. Using calculated tables to produce an aggregated version of a sales table can provide considerable performance gains for any calculation using the aggregated version.

The CALENDARAUTO Function

A handy DAX function that generates a calculated table without using an existing table is the CALENDARAUTO function. This function returns a table with a single column called [Date]. When called, the CALENDARAUTO function inspects every table in the model looking for columns that use either the Date or DateTime datatype.

The oldest and newest date values that appear in any column using these datatypes are used to generate a row for every day between the oldest and newest values. The dates are rounded to the start and end of the calendar year.

Date tables are invaluable for data models, particularly when you add measures designed to show period comparison and running totals accurately. There are several DAX functions that allow you to add time-intelligence logic to your model. These often rely on a date table that has contiguous dates to work properly. In the WideWorldImportersDW dataset, no rows exist in the 'Fact Sale' table with an [Invoice Date Key] that falls on a

Sunday. This might cause problems for some functions when they are processing period comparison logic. Date/Calendar tables help make calculations behave more reliably when there are gaps in dates in non-date table data.

Once you create a calculated table using the CALENDARAUTO function as shown in Figure 1-8, you can start adding calculated columns and measures to it. You can also start creating one-to-many relationships to other tables in your model.

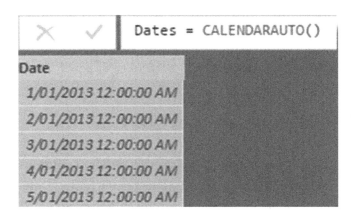

Figure 1-8. *A sample output of the CALENDARAUTO function*

Datatypes

In DAX, it is possible to define datatypes for individual columns. The different datatypes are listed momentarily in Table 1-2 and fall into three main categories, with the exception of True/False (Boolean).

- Text
- Numeric
- DateTime

By select the best datatype, you help reduce the size of your model as well as improve performance when refreshing data and using your report.

When importing new data, the data modeling engine guesses at what the datatype for each column should be. This is something worth checking on as there may be opportunities to adjust to make sure that appropriate data types are used for each column.

Another factor to be aware of is that although your data source may produce numeric data at the time you author your model, at some point in the future, it may start to include nonnumeric data that will cause errors during a data refresh. This is more likely to happen when a data source provides few suggestions as to the best datatype (for example, CSV format).

Table 1-2. *Datatypes in DAX*

Datatype	Stored As	Valid Values
Whole Number	64-bit integer	Signed integer between −9,223,372,036,854, 755,808 and + 9,223,372,036,854,755,808.
Decimal Number	64-bit real number	A mixture of real and decimal numbers ranging between −1.79E + 308 and −2.23E −308. Only 15 significant digits return a precise result. Values outside these calculate and return approximate results.
Fixed Decimal/Currency	64-bit real number	The same as for Decimal Number but fixed to four decimal places.
Date/Time	64-bit double-floating	1st January 100AD to 31st December 9999AD.
True/False		
Text	Unicode (two bytes per character)	The maximum string length is 268,435,456 Unicode characters.

The Whole Number Datatype

The Whole Number datatype, as the name suggests, allows you to store and manipulate data in the form of positive or negative integers. Any decimal component is removed and arithmetic using whole number values returns a whole number and not a value of another datatype.

```
VAR WholeNumberA = 3
VAR WholeNumberB = 2
RETURN DIVIDE( WholeNumberA, WholeNumberB ) // will be stored as 1.5
```

Note   When performing a division calculation, use the DIVIDE function rather than the / operator. Use DIVIDE (3, 2) instead of 3 / 2. This handles cases with a divide-by-zero error more elegantly.

The Decimal and Fixed Decimal Number Datatype

Decimal numbers use 64 bits of storage per value and can represent a wide range of values that either represent a large integer component with a small decimal component or vice versa.

You can store extremely large or small numbers using this datatype and then perform calculations. However, when using this datatype, be aware that with extremely large values, results can be imprecise (not exact).

The following exercise shows values that are stored but not displayed, except for the 15 most significant digits.

Calculations on REAL Numbers

The first example creates a calculated measure using the following code:

```
Measure1 = 100000000000000000 + 50
```

In this code, 17 zeros following the 1 and the calculation returns the value of 100000000000000000. Note that nothing has changed despite the addition of the value of 50. I understand that the considerable number of zeros might make you go cross-eyed, but let me assure that there are 17 zeros.

Now create a second calculated measure that makes a reference to the first as follows:

```
Measure2 = [Measure1] + 50
```

This returns 100000000000000100—a 1, 14 zeros, a 1, and then two more zeros.

If your number has a large integer component, such as the top rows shown in Figure 1-9, DAX allows less precision on the fractional side. The maximum number of digits total on either side of the decimal point is 15.

Decimal Numbers in DAX
123,456,789,012,345.000000000000000
12,345,678,901,234.100000000000000
1,234,567,890,123.120000000000000
123,456,789,012.123000000000000
12,345,678,901.123500000000000
1,234,567,890.123460000000000
123,456,789.123457000000000
12,345,678.123456800000000
1,234,567.123456790000000
123,456.123456789000000
12,345.123456789000000
1,234.123456789010000
123.123456789012000
12.123456789012300
1.123456789012350
0.123456789012346

Figure 1-9. Data formatting

The integer side takes priority, so if you have a value with 12 integer places and 12 decimal places, DAX will automatically round the decimal to 3 places.

Fixed Decimal (or Currency) uses the Decimal datatype, however the fractional component is locked to 4 places. You should show preference for this datatype when you are working with money or currency data. Arithmetic performed using this datatype should yield results that are acceptable for financial reports, particularly regarding rounding and truncation.

Date and DateTime

There are two date-based datatypes available to choose from when you are working with date-based data. The difference between Date and DateTime is, as the names suggest, that DateTime can represent values that include hour, minute, second, and millisecond, whereas Date does not.

It is possible to compare and match values that are both Date and DateTime, however, only values of DateTime that are midnight can match a value of Date from the same day. This can sometimes trip up inexperienced users when they are creating relationships between tables in which one side of the relationship uses a Date datatype while the other uses DateTime. If the table using DateTime has a value such as '2018-01-01 10:30:00', it never finds a matching record in the table using the Date datatype. No error is thrown.

If you need to generate a specific date in DAX, you can use the DATE function, which takes three parameters (year, month, day). DATE(2018, 5, 1) returns a date that represents May 1, 2018.

Time

You can use the Time datatype to represent specific points in time in a day. These values can be "2:37 pm" or "10:30 am" and can be added or combined to make time-based calculations.

```
TIME(1,0,0) + "03:00:00"
```

The preceding calculation shows two different notations of a Time value being added. It is adding 1 hour to a value that represents 3 am. The result is4 am.

Values that use the Time datatype are still stored as a full DateTime and use December 30, 1899, for the year, month, and day components. Functions that use Time typically ignore the year, month, and day.

Operators

These are the four sets of operators in DAX:

- Arithmetic
- Comparison
- Concatenation
- Logical

There are no bitwise operators in DAX. Bitwise calculations can be performed using functions rather than operators.

Arithmetic Operators

Table 1-3 shows the arithmetic operators available in DAX.

Table 1-3. *DAX Atrithmetic Operators*

Operator	Effect	Example
+	Addition	2 + 2 = 4
–	Subtraction	10 – 4 = 6
*	Multiplication	4 * 5 = 20
/	Division	8 / 2 = 4
^	Exponents	2 ^ 4 = 16

Comparison Operators

The operators in Table 1-4 return true or false when used to compare two values. Values on either side of the operator can be Text, Numeric, or DateTime.

Table 1-4. *DAX Comparison Operators*

Operator	Effect	Example
=	Equal to	Sales[Qty] = 10
<	Less than	Sales[Qty] < 10
>	Greater than	Sales[Price] > 20
<=	Less than or equal to	Dates[Date] <= TODAY ()
>=	Greater than or equal to	[Total] >= 200
<>	Not equal to	[Animal] <> "Cat"

If you compare a number to a text value, for example, 1 = "1", you receive an error. In this case, you can convert the numeric 1 to a string using the FORMAT function, or you can convert the "1" value to a number using the VALUE function.

Concatenation Operator

The operator used for concatenating text values is shown (Table 1-5).

Table 1-5. *DAX Concatenation Operator*

Operator	Effect	Examples
&	Concatenates values	"ABC" & "DEF" = "ABCDEF"
		1 & 2 = "12"

In the "1 & 2" example, both values are implicitly converted to text with the output being a text value.

Logical Operators

Table 1-6 show logical operators in DAX.

Table 1-6. *DAX Logical Operators*

Operator	Effect	Example
&&	Logical AND	(1=1) && (2=2) = true
		((1=1) && (2=3) = false
||	Logical OR	(1=1) || (2=2) = true
		(1=1) || (2=3) = true
		(1=2) || (2=3) = false
IN	Logical OR	Location[Country] IN ("UK","USA","Canada")

Operator Precedence

Once you have the correct operator, you also need to consider the order in which to apply operators. Table 1-7 shows the order of operator precedence in DAX.

Table 1-7. *Operator Precedence*

Operator	Effect
^	Exponent
−	Sign (positive or negative)
* /	Multiplication and division
!	NOT
+ −	Addition and subtraction
&	Concatenation
= < > <= >= <>	Comparison

You can use parentheses to override the operator precedence if you need to as shown in the examples in Table 1-8. In the example on the second row, the addition takes place before the multiplication, producing a different result than the first row. In the example on the bottom row, the parentheses make the sign operator take precedence over the exponent operator.

Table 1-8. *Examples of Operator Precedence*

Calculation	Result
2 * 3 + 4	10
2 * (3 + 4)	14
−2 ^ 2	−4
(−2) ^ 2	4

DAX attempts to implicitly convert two sides of an operator to the same datatype where possible. This cannot be guaranteed in all cases. So, for instance, 1 + "2" results in an error whereas 1 & "2" results in the text value of "12".

Relationships

Relationships are rules in DAX that define how two tables can be associated. There are two main reasons to create relationships. The first is to allow filter selections made on one table to automatically filter rows on another table via the relationship. The other is to allow calculations to use values from rows in different tables and to understand how rows should be connected.

Types of Relationships

Relationships can be defined as *one-to-many* or *one-to-one* but always use a single column from each table in the relationship. When you add a relationship between two tables, the column involved on one side of the relationship must have unique values. If duplicate data is detected during a data load or refresh, an error occurs telling you that you can't create a relationship between the two columns because the columns must have unique values.

Relationships for Filtering

The WideWorldImportersDW dataset contains several fact tables that each have one or more date-based columns. The dataset also contains a table called Dimension.Date, which has one row per day between January 1, 2013, and December 31, 2016. The date table contains 14 columns useful for grouping and filtering.

Once you import these tables to the model, you can add relationships between pairs of tables that allow report-based filters to make any selection on the table on the one side flow through and filter rows on the related table.

Filters only flow across relationships in one direction. A filter selection made to a column in a table on the many side of a relationship will not filter rows in the table on the one side. Filters can flow down through multiple sets of relationships, but always in the direction of one to many.

This means that if you add any field from the date table to a report visual or filter, any selection you make using these visuals or filters propagates to related tables and triggers measures to recalculate using the new filter settings.

Figure 1-10 shows a relationship between the 'Dimension Date' table and the 'Fact Sale' table. The column used on the 'Dimension Date' (one) side is [Date], whereas the column used on the 'Fact Sale' (many) side is [Invoice Date Key]. Both columns use the Date datatype.

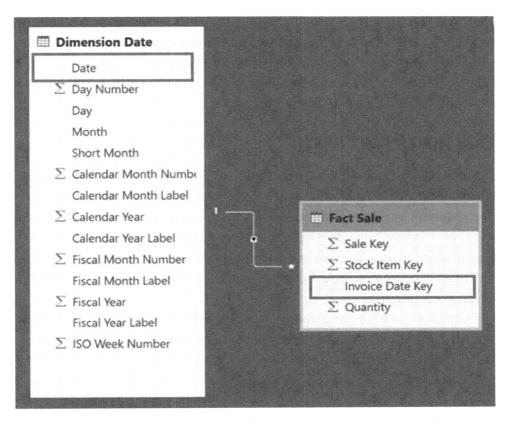

Figure 1-10. *The relationship between 'Dimension Date' and 'Fact Sale'*

With this relationship in effect, drag the [Calendar Year] column from 'Dimension Date' to the report canvas and turn it into a slicer visual. The slicer shows distinct values from the column, which are 2013, 2014, 2015, and 2016. Any selection of one or more of these values flows down and is applied to any related table. If 2014 is selected in the slicer, calculations used by visuals based on columns in related tables automatically recompute and produce new values that consider the updated filtering on the 'Dimension Date' table.

Power BI Desktop has a feature that can automatically detect relationships when you add new tables to your model. If a new table contains a column that shares the same name and datatype to a column in another table, a relationship may be added for you. This can be helpful, but it is always good to double check what relationships have been added for you and if these are indeed useful.

A good tip I picked up from my good friend Matt Allington is to organize tables in the Relationship View so tables on the one side are positioned higher than tables on the many side. This helps visually reinforce the trickle-down nature of filtering through relationships. It becomes harder when you have many tables in your data model, but it's a tip well worth applying for smaller models.

Relationships in Calculations

The other way you can take advantage of relationships is through the DAX functions RELATED and RELATEDTABLE. These functions allow calculations to use data from rows from different, but related tables. This means you can add a calculated column to a table that shows a value using data from a parent or child table. The RELATED function allows calculations to tunnel up to data in a table on the one side from the many side, whereas RELATEDTABLE provides access to data from a table on the many side to a calculation on the one side.

A pair of tables can have more than one relationship defined, but only one can be marked as the active relationship at any one time. Active relationships are used by default by filters and calculations. Relationships that are not active can be used in calculations, but they need to be specifically referenced using the USERELATIONSHIP filter function.

Although it is possible to make your table/relationship mirror that of an OLTP, it is far better to study and learn some BI methodology, such as using a star schema to organize your data. This is especially useful once the row counts in your tables grow to large values.

It's also possible to define multiple relationships between two tables and control the cross-filter direction. These concepts are covered in more detail in the chapter on DAX relationships (Chapter 5).

Hierarchies

Hierarchies combine two or more columns from a table together as a level-based grouping. These can be added to your model by dragging a field and dropping it on another field from the same table. This creates a new entry in the field list with a special icon to show it is a hierarchy. The fields that represent the levels in the hierarchy appear slightly indented and can be reordered by dragging and dropping.

You can rename a hierarchy as well as individual levels. Renaming these has no impact on the columns used as source columns for the hierarchy.

In the example in Figure 1-11, the [Calendar Month Label] field has been dragged and dropped onto the [Calendar Year] field, which creates a new hierarchy. The two levels and the hierarchy are then renamed, making each name shorter. Finally, a third level is added to the hierarchy by the [Day] field being dragged and dropped onto the hierarchy name. This adds the field as a new third level. It's possible to reorder by dragging individual levels up or down the list.

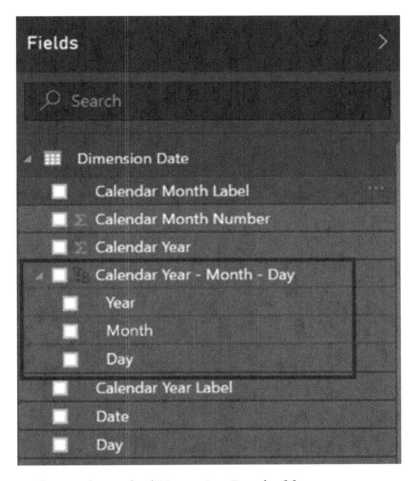

Figure 1-11. *A hierarchy in the 'Dimension Date' table*

Not all Power BI visuals understand hierarchies, but most of the native visuals allow hierarchies to be used on an X or Y axis to provide drill-down functionality. You can obtain the same experience by dropping multiple individual fields to the same axis fields, but you have less ability to rename and customize the levels.

CHAPTER 2

Variables

This chapter looks at how you can use DAX variables to make calculations easier to understand and, in some cases, perform faster. Variables are used to store results from DAX expressions. These variables can be as simple as assigning a hardcoded value or the result of a complex DAX equation.

You can use variables in any type of DAX calculation including calculated columns, measures, and tables. Variables are not strictly typed and can represent any type of object in DAX. Variables automatically assume the type of the object being assigned. You must use a RETURN statement to close a layer of variable scope. You can declare multiple variables within the same layer of scope and you can use them with a single RETURN statement. Nested variables can initialize a new layer of scope when you use them anywhere within an expression, but all layers of scope are ended when you use a RETURN statement.

This is the basic structure of creating and using a variable in your DAX expression:

```
VAR varname = expression
RETURN expression
```

You can only assign variables once and cannot reassign them. The following code produces an error:

```
VAR myVar = 1
myVar = myVar + 1
```

The following example shows the basic structure of a formula using multiple DAX variables:

```
VAR myVar1 = 1
VAR myVar2 = myVar1 + 2
RETURN myVar2 * 2
```

27

© Philip Seamark 2018
P. Seamark, *Beginning DAX with Power BI*, https://doi.org/10.1007/978-1-4842-3477-8_2

Variable Structure

The keyword VAR denotes a new variable being declared. This is followed by the name of the variable. You cannot use spaces or encapsulate the variable name with square brackets or apostrophes to allow spaces like you can elsewhere in DAX. Another limitation on variable names is that you cannot use the names of tables or DAX keywords.

Finally, a DAX expression follows the = operator.

```
VAR myVar1 = 1
VAR myVar2 = myVar1 + 2
```

The first line of this example assigns the numeric value of 1 to a variable called myVar. The myVar variable is used in an expression on the following line to assign a value to the variable called myVar2. At this point, the variable myVar2 carries the number 3 because of the expression (1 + 2).

```
RETURN myVar2 * 2
```

The RETURN keyword must be used to output the result and can also use an expression. In this case, the result is the whole number 6. The RETURN keyword can only be used once per variable scope.

Note   The RETURN keyword is used to return the value of any variable in the current scope. This can be useful when you're debugging calculations with multiple variables. It does not have to return the last variable in the series.

A major benefit of using variables in DAX is code readability. In plenty of cases, you can simplify complex DAX formulas by adding variables. The following example shows an unformatted DAX expression that doesn't use variables.

```
Table =
    NATURALINNERJOIN(
    'Sales',
    TOPN(
      10,
      SUMMARIZECOLUMNS(
        'Sales'[Product],
        "Sum of Revenue", SUM('Sales'[Total])
        ),
    [Sum of Revenue],
    DESC))
```

In the following code, the DAX expression is now formatted and rewritten to make use of variables. It produces the same result as the preceding code.

```
Table =
VAR InnerGroup =
    SUMMARIZECOLUMNS(
        -- Group BY --
        'Sales'[Product],
        -- Aggregation Column --
        "Sum of Revenue", SUM('Sales'[Total])
        )

VAR Top10PRoducts =
    TOPN(
        10,
        InnerGroup,
        [Sum of Revenue],
        DESC
        )

RETURN
    NATURALINNERJOIN(
        'Sales',
        Top10PRoducts
        )
```

Although this example is longer, breaking the code into smaller steps and formatting it makes it much easier to read and understand its logic.

Using Variables with Text

Variables can store text as well as numeric data. The variable inherits the type of the expression being assigned. This example joins the text from textVar1 and textVar2 to produce a new value that returns "Hello World".

```
My Measure =
VAR textVar1 = "Hello "
VAR textVar2 = "World"
RETURN CONCATENATE(textVar1,textVar2)
```

You can use this technique with other DAX functions to create dynamic text-based measures that may change depending on the current filter context. Consider the following calculated measure:

```
Sales Text =
VAR SalesQty = SUM('Fact Sale'[Quantity])
VAR Text1 = "This month we sold "
VAR Text2 = " Items"
VAR Result = IF(
                SalesQty > 0,
                -- THEN --
                Text1 & SalesQty & Text2,
                -- ELSE --
                "No sales this month"
                )
RETURN Result
```

Here you assign the DAX SUM expression to the SalesQty variable. This is used later in the IF function to test for the existence of sales to determine the output message. Note the use of comments in the IF function to help clarify which code is being used to update the Result variable.

This also shows how you can use variables to improve performance. The SalesQty variable is potentially used twice in the IF function. If this formula was written without using variables, the IF function would make two calls to the underlying column and therefore take longer to arrive at the same result.

A similar function using text-based variables is one that generates a greeting calculated measure. Creating a calculated measure using the code in Listing 2-1 tests the current time of day and stores the hour of the day in the CurrentHour variable. This is then used in a SWITCH function to generate appropriate text to be stored in the GreetingText variable. This is then combined with other text to produce a measure that you can use as a dynamic greeting on your report.

Listing 2-1. Generating a Greeting Calculated Measure

```
Greeting =
VAR CurrentHour = HOUR(NOW())
VAR GreetingText =
    SWITCH(
        TRUE(),
        CurrentHour<12,"Morning",
        CurrentHour<17,"Afternoon",
        "Evening"
        )
RETURN
    "Good " & GreetingText & ", " & USERNAME()
```

Using Variables in Calculated Columns

When you are using variables inside calculated columns, variables automatically have access to values in any column from the same row. When you use the 'tablename'[columnname] notation during the assignment, the assumption is that the value used is from the same row.

This example uses variables to help create a new column that combines separate location values into a single piece of text in the new column. This type of column can be useful for some visuals that try to plot using text-based address data.

In this case, the DAX expression for the calculated column might look like this:

```
City and Country =
VAR city = 'Dimension City'[City]
VAR country = 'Dimension City'[Country]
RETURN
    city & ", " & country
```

In this block of code, you have declared two variables, city and country. The datatype inherited for these variables is Text and uses the & operator to concatenate the variables together in the RETURN statement. A sample of the output once the calculated column has been added is shown in Figure 2-1.

```
City and Country =
VAR city = 'Dimension City'[City]
VAR country = 'Dimension City'[Country]
RETURN
    city & ", " & country
```

WWI City ID	City	State Province	Country	City and Country
11945	Fort Douglas	Arkansas	United States	Fort Douglas, United States
12112	Fox	Arkansas	United States	Fox, United States
12518	Gainesville	Arkansas	United States	Gainesville, United States
12581	Gamaliel	Arkansas	United States	Gamaliel, United States
12834	Genoa	Arkansas	United States	Genoa, United States
12957	Gifford	Arkansas	United States	Gifford, United States
13162	Glencoe	Arkansas	United States	Glencoe, United States
13179	Glendale	Arkansas	United States	Glendale, United States
13419	Goodwin	Arkansas	United States	Goodwin, United States
13577	Grand Glaise	Arkansas	United States	Grand Glaise, United States

Figure 2-1. *Sample output of new calculated column using variables*

The result of the preceding code is a new column added to the table called [City and Country]. The values for each row are the result of concatenating text using other values from the same row. This is similar to what you'd expect when you are creating new columns in a table in Excel, or when you are creating a new column in a SELECT statement in SQL based on other values from the same row.

When you use variables in calculated columns, the final RETURN statement must return a single value. It cannot return a column or a table.

Using Variables in Calculated Measures

You can use variables to make calculations in calculated measures more readable and, in some cases, perform faster.

A difference between variables used in calculated measures compared with calculated columns is that variables in calculated measures do not have a connection with individual rows like they do with calculated columns. This means variables used in calculated measure cannot be assigned a column-based value.

Using the same DAX code as in the previous section, a new calculated measure called [City and Country] would encounter errors similar to this:

```
A single value for column 'City' in table 'Dimension City' cannot be determined
```

This is more the nature of the difference between calculated measures and calculated columns, which are covered in Chapter 9. So, for now, let's create a calculated measure using variables that work.

Say you have a report requirement to understand how many sales have a [Total Including Tax] value that is greater than or equal to $100.

One way to write this as a calculated measure using variables shown in Listing 2-2.

Listing 2-2. Writing Code as a Calculated Measure Using Variables

```
Total Sales > $100 =
VAR SalesAbove100 =
            FILTER(
                    -- Filter Table --
                    'Fact Sale',
                    -- Filter Condition --
                    'Fact Sale'[Total Including Tax]>=100
                    )
RETURN COUNTROWS(SalesAbove100)
```

In this calculated measure, a variable called SalesAbove100 is used to assign the output of the FILTER function. The FILTER function returns a table that cannot be used as the final output for the RETURN function because you are defining a calculated measure and calculated measures must return a single value.

The RETURN statement uses the COUNTROWS function to return a single value based on the table passed to it as the argument. This calculated measure shows how you can use the COUNTROWS function over a table expression stored in a variable and that is not limited to physical tables in the model.

Using Variables in Calculated Tables

You can also use variables in DAX calculations to create new calculated tables. You may find this useful when you are creating new tables as summary or aggregated versions of existing tables that require additional processing or manipulation.

This technique can improve the performance of some report pages, although it will have an impact on the time it takes to load data as well as the overall memory footprint of your data model.

The example in Listing 2-3 uses variables to help create a table showing who the top ten customers from 'Fact Sale' are based on amounts in the [Total Including Tax] column.

Listing 2-3. Using Variables to Create a Table

```
Sales Summary =
VAR SalesTableWithCustKey =
     FILTER(
                    -- Filter Table --
                    'Fact Sale',
                    -- Fitler Condition --
                    'Fact Sale'[Customer Key]>0
                    )
VAR SalesTableGrouped =
     SUMMARIZE(
                        -- Table to Summarize --
                    SalesTableWithCustKey,
                    -- Columns to Group By --
                    'Fact Sale'[Customer Key],
                    -- Aggretated Column Name & Expression --
                    "Total Spend",SUM('Fact Sale'[Total Including Tax])
                        )
VAR SummaryTableFiltered =
     TOPN(
             -- Number of rows to return
                    10,
                    -- Table to filter --
                    SalesTableGrouped,
                    -- Order by Expression
                    [Total Spend]
                    )
RETURN SummaryTableFiltered
```

The first variable in the DAX calculation is SalesTableWithCustKey, which is assigned the output of the FILTER function used to remove rows of data that have a [Customer Key] of zero.

The SalesTableGrouped variable is assigned a table that is the output of the SUMMARIZE function. The SUMMARIZE function takes the SalesTableWithCustKey variable and groups by the [Customer Key] column. A new column called [Total Spend] is added, which calculates the SUM values for all rows belonging to each [Customer Key] to one value.

Finally, the SummaryTableFiltered variable is assigned the output of the TOPN filter function that filters a table with all customers down to a table showing just the top ten. This variable is eventually returned to the calculated table (Figure 2-2).

Customer Key	Total Spend ↓
149	438689.81
132	427445.57
381	420001.08
14	414934.18
273	408524.95
394	405824.67
351	404945.65
311	403635.05
370	398563.46
102	395482.71

Figure 2-2. *Sample output of the 'Sales Summary' calculated table*

Unlike with the calculated column and calculated measure examples, you can return a calculated table to the data model when you're using calculated tables.

The equivalent T-SQL using #tablenames that match the variables for the earlier DAX at Listing 2-3 would look like Listing 2-4.

Listing 2-4. Equivalent T-SQL to Match the DAX Variables

```
SELECT
    *
INTO #SalesTableWithCustKey
FROM Fact.Sale
WHERE
    [Customer Key] > 0

SELECT
    [Customer Key],
    SUM([Total Including Tax]) AS [Total Spend]
INTO #SalesTableGrouped
FROM #SalesTableWithCustKey
GROUP BY
      [Customer Key]

SELECT TOP 10
    *
INTO #SummaryTableFiltered
FROM #SalesTableGrouped
ORDER BY
      [Total Spend] DESC

SELECT * FROM #SummaryTableFiltered
```

Debugging Using Variables

A common technique used in SQL Stored Procedures is to create a series of temporary tables that are refined versions of a previous table. You can use variables in DAX in an equivalent way that can make it easier to follow and debug the processing of your data.

The final RETURN statement does not always have to return the most recent variable. It can be useful to output any variable so you can visually inspect the values contained by that variable. This is a form of manual debugging.

Nesting Variables

Variables can be nested and multiple layers of variable scope can exist within the same calculation.

Each layer of variable scope begins with a VAR statement and ends with a matching RETURN statement and can only reference other variables declared in the same level or higher.

Listing 2-5 is a simple example of a calculated measure that returns the value of 30. Two layers of variable scope are being used in this calculation.

Listing 2-5. An Example of a Calculated Measure

```
Nested Measure =
VAR Level1a = 10
VAR Level1b =
        VAR level2a = Level1a
        VAR Leval2b = level2a * 3
        RETURN Leval2b
RETURN Level1b
```

The first VAR statement begins the outermost layer of variable scope. The highlighted code shows the start and finish of the inner layer of variable scope and ends by returning a single value to be assigned to the Level1b variable.

The code in the inner layer of scope can access variables that have been created in the outer layer. This is shown by the Level2a variable being assigned the same value as Level1a. This cannot happen the other way around. Level2a should have a value of 10. Level2b should have a value of 30.

Listing 2-6 shows a slightly more complex version with multiple levels of nesting.

Listing 2-6. Complex Version with Multiple Nested Levels

```
Nested Measure 2 =
VAR Level1a = 10
VAR Level1b =
        VAR level2a = 20
        VAR level2b =
```

```
            VAR Leve3a = Level2a + Level1a
            RETURN Level1a
        RETURN level2b
VAR Level1c =
        VAR Level4a =  Level1b * 5
        RETURN Level4a

RETURN Level1c
```

This calculated measure returns a value of 50. There are three layers of nesting below the Level1b variable. The innermost layer can access variables declared at all higher levels. The innermost level doesn't return a variable declared in the same level.

The layer of scope opened beneath the Level1c variable cannot access any variables declared by the layers that were opened (and closed) during the expression to assign a value to Level1b.

Nested Variables (Complex)

Rather than using hardcoded values to demonstrate variable nesting, Listing 2-7 shows how you can use variables in other types of calculations.

Listing 2-7. Example with Variables in Other Types of Calculations

```
Demo Table =

VAR Level1 =
    SUMMARIZE(
        FILTER(
            'Fact Sale',
            [Quantity] >
                    VAR Level2 = DAY(TODAY())
                    RETURN IF(
                            Level2 > 10,
                            -- Then --
                             20,
                            -- Else --
                            30
                            )
```

```
        ),
        -- Group by Columns --
        [City],
        -- Aggregations Columns --
        "Sum of Quantity",sum('Fact Sale'[Quantity]))
RETURN
    TOPN(
        VAR Level2 = 5
        RETURN Level2,
        Level1,
        [City]
        )
```

This calculated table uses nested variables to help control the parameters used in the filter expression of the FILTER function. The Level2 variable can be assigned different values depending on the day of the month. If the current day of the month is between 1 and 10, this layer of scope returns a value of 30, otherwise it returns a value of 20 to be used in the filter expression.

Another nested level of scope that safely uses the same name previously is assigned a hardcoded value of 5 to be returned as a parameter to the TOPN function that also uses the Level1 variable as a parameter.

This is not the most useful calculation, but it does demonstrate how you can use variables throughout longer calculations.

CHAPTER 3

Context

Some say that once you have mastered the concept of context in DAX, you have mastered DAX. This is mostly true, and if you have indeed mastered context, you are a long way up the learning curve.

Context can seem a little daunting at first, but once you understand the effect that types of context have on calculations, hopefully DAX starts to make more sense. *Context* is how DAX applies layers of filtering to tables used in your calculations so that they return results that are relevant for every value.

Most context is automatic, but some context allows you to control the underlying data passed to your DAX calculations. This chapter mostly uses pivot tables to describe the order and effect of the different types of context. Other visuals and charts apply the context logic the same way pivot tables do, but I think pivot tables do a good job of showing the effect.

There are two types of context in DAX: filter context and row context. Depending on your calculation, one or both can apply at the same time to affect the result of your DAX calculation.

Context is the layer of filtering that is applied to calculations, often dynamically to produce a result specific to every value in a pivot table or visual, including row and column totals.

Filter Context

Filter context is a set of column-based filters applied on the fly to the underlying data for every DAX calculation. It is helpful to think of every value cell in a pivot table as its own separate computation. Each value cell treats its measure as a function and passes relevant filters to the function to be applied to the calculation.

© Philip Seamark 2018
P. Seamark, *Beginning DAX with Power BI*, https://doi.org/10.1007/978-1-4842-3477-8_3

It might also be helpful to think of filter context as a container. The container can be empty or have one or more column filter rules. A column filter rule simply specifies which rows in a given column in a table return true to a Boolean test.

Although examples in this chapter are mostly single-column filters, you should not assume that all filters in a filter context are single-column filters. In general, each filter in the filter context is a table that can have more than one column. This is important because a two-column filter establishes a correlation between the two columns. For example, a filter of two rows { (2016, US), (2017, UK) } is different from two independent column filters { 2016, 2017 } and { US, UK }.

Each execution of a calculation starts by having access to every row of every table in the data model, but before the core calculation is executed, a filter context is established to determine the most relevant set of column filters for that specific execution. Once the filter context is finalized, the column filter rules contained inside the context are applied across the data and calculations such as SUM execute using remaining data in the columns they have been coded to use. Once the calculation is complete, the filter context is destroyed and is not used by any other process.

Column-based filters can be added to the filter context implicitly, in that they are added automatically to the filter context because of another field in the same visual, such as a row or column header. They can also be added to the filter context using filters external to the visual, such as slicer selections or other report, page, or visual filters.

Column-based filters can also be added to, or removed from, the filter context explicitly. This is where filter-based rules are added into DAX calculations.

Unlike T-SQL, in which you must write specific code to filter and aggregate using JOIN, WHERE, GROUP BY, and HAVING clauses in SELECT statements, DAX can dynamically construct predicate conditions per computation for you.

Let's start with a very basic DAX measure that uses the SUM function to return a value that represents all the values added together for an individual column.

```
Sales Qty = SUM('Fact Sale'[Quantity])
```

This calculated measure returns a number that represents the sum of every value in the [Quantity] column in the 'Fact Sale' table. Dropping this calculated measure onto a suitable visual on a report with no additional fields or filters applied should display a value of 8,950,628 using the WideWorldImportersDW dataset.

In this example, the single value displayed on the report represents a single execution of the calculated measure. The [Sales Qty] calculated measure uses the SUM function that is passed to the column reference of 'Fact Sale'[Quantity], meaning it adds every value from every row in the [Quantity] column to produce a value. It has access to every value in the [Quantity] column because there are no column-based filters in the filter context.

The following would be the equivalent T-SQL statement to this DAX:

```
SELECT
    SUM(Quantity) AS [Sales Qty]
FROM Fact.Sale;
```

Implicit Filter Context

When a calculated measure is added to a pivot table using other fields from 'Fact Sale' (or from tables related to 'Fact Sale') in the rows and columns, we don't see the value 8,950,628 repeated over and over in every cell; instead we see other values that represent the calculation filtered according to the intersecting points of our column and row headers (Figure 3-1).

This is the effect of filter context on the calculated measure. DAX is implicitly adding a set of column-based filters to the filter context for every one of the 20 cells in the pivot table, and a different result is showing in every value cell. This pivot table is using the simple [Sales Qty] calculation that contains no DAX code to say how to group or aggregate the [Quantity] column to produce each value.

Calendar Year	Far West	Mideast	Southeast	Total
2013	277,812	363,441	531,784	1,173,037
2014	298,948	386,863	562,337	1,248,148
2015	314,145	399,124	606,844	1,320,113
2016	137,765	187,404	268,302	593,471
Total	1,028,670	1,336,832	1,969,267	4,334,769

Figure 3-1. *The [Sales Qty] calculated measure in a pivot table*

Consider Figure 3-1, which uses the [Sales Qty] calculated measure. The pivot table has five rows and four columns. There are four calendar years in the dataset, which is why you are seeing five rows (including a row at the bottom for the total). The Sales Territory has been filtered to show just three of the ten actual Sales Territories in the data for this example. The right-most column is a grand total for each row.

The values in each cell are different and none are the same as the grand total value of 8,950,628. Every execution of the calculated measure in this pivot table uses a unique filter context.

The DAX expression is a simple SUM calculation and makes no mention of any filtering by Calendar Year, or Sales Territory region, or any other grouping instruction for that matter. So why do we have the value of 277,812 in the top-left cell, and 298,948 in the cell immediately below that?

This is the effect of filter context in DAX. The pivot table has 20 value cells to be calculated, so the DAX engine performs 20 separate logical calculations, one logical computation for every cell in the pivot table. This includes the calculations required to determine the values for each of the row and column totals.

To produce a value of 277,812 for the top-left cell, the calculated measure begins by having access to every value in the 'Fact Sale'[Quantity] column. The filter context for this calculation starts empty but has two sets of column filters implicitly added to it. One of the column filters is that rows in the [Calendar Year] field must have a value of 2013; the other column filter is that rows in the [Sales Territory] column must have a value of "Far West".

All filters in a filter context are based on a logical AND, so only rows in the 'Fact Sale'[Quantity] column that can meet both criteria are passed to the SUM function.

The T-SQL equivalent to generate the value for this cell is like adding the INNER JOIN and WHERE clauses in Listing 3-1 to the T-SQL statement.

Listing 3-1. T-SQL of the [Sales Qty] Calculated Measure with Filters to Produce Value for Top-Left Cell in Figure 3-1

```
SELECT
    SUM([Quantity]) AS [Sales Qty]

FROM FACT.Sale AS S

    INNER JOIN Dimension.City AS C
        ON C.[City Key] = S.[City Key]
```

```
      INNER JOIN Dimension.Date AS D
            ON D.Date = S.[Invoice Date Key]
WHERE
      C.[Sales Territory] = 'Far West'
      AND D.[Calendar Year] = 2013
```

The WHERE clause in T-SQL is like the filter context. The two statements inside the WHERE clause represent each of the column filters that were implicitly added to the filter context.

Let's now look at what the DAX engine is doing to produce the value for the second row of the first column (298,948). The filter context for this computation again starts empty but has two sets of column filters implicitly added to it. One column filter is that rows in the [Calendar Year] field must now have a value of 2014; the other column filter is that rows in the [Sales Territory] column must have a value of "Far West". Listing 3-2 shows the equivalent T-SQL for the computation used for this cell.

Listing 3-2. T-SQL of the [Sales Qty] Calculated Measure with Different Filter Context

```
SELECT
      SUM([Quantity]) AS [Sales Qty]
FROM FACT.Sale AS S

      INNER JOIN Dimension.City AS C
            ON C.[City Key] = S.[City Key]

      INNER JOIN Dimension.Date AS D
            ON D.Date = S.[Invoice Date Key]
WHERE
      C.[Sales Territory] = 'Far West'
      AND D.[Calendar Year] = 2014
```

The only difference between the T-SQL statements in Listing 3-1 and Listing 3-2 is the final predicate that specifies that the value for D.[Calendar Year] is now 2014 rather than 2013. Each cell has its own filter context and each filter context has its own set of column filters.

This is repeated over and over in any order because no cell in a pivot table relies on the output of another cell.

Let's jump to the bottom row of the first column and look at what the DAX engine is doing to produce a value of 1,028,670 for the grand total of the first column. DAX does not keep a running total of the previous calculations in the same column. Logically the engine starts from scratch and arrives at a value for the total independent of the other 19 cell calculations in the pivot table.

DAX once again starts with an empty filter context but in this case, just one column filter is implicitly added to the filter context. This time the column filter is that the [Sales Territory] column must have a value of "Far West". This filter context restricts the rows in the 'Fact Sales'[Quantity] column that are passed to the SUM function. More rows are passed in this case than from the previous two examples.

Listing 3-3 shows the T-SQL code for this cell.

Listing 3-3. T-SQL Equivalent for Basic DAX Query to Produce the Total Value for the First Column of Figure 3-1

```
SELECT
      SUM([Quantity]) AS [Sales Qty]
FROM FACT.Sale AS S

      INNER JOIN Dimension.City AS C
            ON C.[City Key] = S.[City Key]
WHERE
      C.[Sales Territory] = 'Far West'
```

The difference between this and the earlier examples is you no longer need an inner join to the Dimension.Date table and you have one less predicate in your WHERE clause, so no filtering is taking place on the date table.

You can use this to your advantage in more complex scenarios where it's possible to specify an altogether different calculation for the grand total cell than the calculation used in the cells filtered by a calendar year. In fact, it's possible to use functions to instruct DAX to return completely different calculations altogether for any cell you want to produce the SCOPE-like behavior available in multidimensional data models.

The calculated measure in Listing 3-4 applies a test to see if any column filter rules are inside the filter context that 'Dimension Date'[Calendar Year] must match 2014. If there are, the SUM function is bypassed and the text value of "overridden" is returned.

Listing 3-4. Using SELECTEDVALUE to Help Override output

```
Sum of Quantity =
    IF(
        SELECTEDVALUE('Dimension Date'[Calendar Year])=2014,
        -- THEN --
        "overridden",
        -- ELSE --
        FORMAT(
            SUM('Fact Sale'[Quantity]),"#,000"
            )
        )
```

Any cell that uses the [Sum of Quantity] calculated measure that has a column filter for [Calendar Year] = 2014 inside its filter context now returns the "overridden" text instead of running the SUM function (Figure 3-2).

Calendar Year	Far West	Mideast	Southeast	Total
2013	277,812	363,441	531,784	1,173,037
2014	overridden	overridden	overridden	**overridden**
2015	314,145	399,124	606,844	1,320,113
2016	137,765	187,404	268,302	593,471
Total	**1,028,670**	**1,336,832**	**1,969,267**	**4,334,769**

Figure 3-2. *Output of the [Sum of Quantity] calculated measure*

Notice the row totals still produce the same totals as before and have not been reduced by the missing values in the 2014 row. This is because the calculations for the row totals have independent filter context and can still use the data belonging to [Calendar Year] = 2014.

Column filters can be implicitly added to the filter context from sources other than row and column headers of a visual. They can be driven from external sources, such as selections on slicers or other visuals, or they can be specific report, page, or visual filter rules.

If a report-level filter specifies that the 'Dimension City'[Country] field must be "United States", this column filter rule is added implicitly to the filter context for every execution of a calculated measure used in the report.

Another Example

Let's look at filter context that uses a simpler dataset. Consider the dataset in Table 3-1.

Table 3-1. *Dataset to Demonstrate Filter Context*

Person	Pet	Value
Andrew	Cat	10
Andrew	Dog	20
Andrew	Dog	30
Bradley	Cat	40
Bradley	Cat	50
Bradley	Dog	60
Charles	Cat	70
Charles	Bird	80
Charles	Dog	90

Using data from Table 3-1 in a pivot table would produce the results shown in Figure 3-3.

Person	Bird	Cat	Dog	Total
Andrew		10	50	60
Bradley		90	60	150
Charles	80	70	90	240
Total	80	170	200	450

Figure 3-3. *Dataset from Table 3-1 used in a pivot table*

Here is the calculated measure used in the pivot table to generate these values:

```
Sum of Value = SUM( Table1[Value] )
```

Let's walk through an example of filter context. Let's start by using the highlighted cell (Figure 3-3) in the top row of the data that shows a value of 50. This is the value that is the intersection of Andrew and Dog.

The calculation starts with an empty filter context that contains no column-based filters. If the filter context remains like this, the SUM function has access to every row in the [Value] column of Table 3-1 and an output 450.

The calculated measure is on a pivot table that has one field on the row header [Person] and one on the column header [Pet], so two column-based filters are added to the filter context before the calculation executes the SUM function. The first column-based filter is that the Table1[Person] must be "Andrew" (see Figure 3-4). This means the filter context now has a column-based filter rule.

Person	Bird	Cat	Dog	**Total**
Andrew		10	50	**60**
Bradley		90	60	**150**
Charles	80	70	90	**240**
Total	**80**	**170**	**200**	**450**

Figure 3-4. *Showing implict column filter rule of "Andrew"*

Table 3-2. *Dataset from Table 3-1 with Filter Applied to [Person] Column*

Person	Pet	Value
Andrew	Cat	10
Andrew	Dog	20
Andrew	Dog	30

The effect of the filter context on the data is that the [Value] column has now been reduced to just three rows of data as shown in Table 3-2, and if left like this, the SUM function will return 60.

* The second column-based filter to be added to the filter context is that the Table1[Pet] must be "Dog". This now means the filter context has two column-based filter rules (see Figure 3-5). This reduces the rows in the Value column to just two rows (Table 3-3).

Person	Bird	Cat	Dog	Total
Andrew		10	50	60
Bradley		90	60	150
Charles	80	70	90	240
Total	80	170	200	450

Figure 3-5. *Showing implict column filter rule of "Andrew" and "Dog"*

Table 3-3. *Dataset from Table 3-1 with Filter Applied to [Person] and [Pet] Columns*

Person	Pet	Value
Andrew	Dog	20
Andrew	Dog	30

The filter context is complete, and the SUM function can now add all the values in the [Value] column. This results in a value of 50, which is the number that was highlighted back in Figure 3-3.

Column-based filters can also be added or removed from the filter context by code in a DAX calculation. These are known as *explicit filter context* and we look at them next.

Explicit Filter Context

Explicit filter context refers to code in calculations that specifically adds or removes column filter rules to and from the filter context. Explicit filter context is applied after implicit and row context and allows you to customize and sometimes completely override the default behavior of your measures in a way the row and query context can't. You can apply filter context using DAX functions such as FILTER.

Several functions allow you to control the data that is ultimately exposed to your calculations and therefore impact the result. These functions, such as ALL, ALLEXCEPT, and ALLSELECTED are covered in Chapter 6, which is dedicated to filter functions. You can use these to apply or ignore any combination of outside filters already in place to provide fine-grained control over what data is used by your DAX expressions.

Let's start with a simple example that uses WorldWideImportersDW data. Let's create a calculation that shows the SUM of [Quantity] but only for stock items that are the color red.

You can approach this in two ways. You can create a calculated column or a calculated measure. Both exhibit different behaviors, which you see here.

The DAX for the *calculated column* might look like this:

```
Filter on Red as Column = CALCULATE(
                          SUM('Fact Sale'[Quantity]),
                          'Dimension Stock Item'[Color]="Red"
                          )
```

Whereas the DAX for the *calculated measure* might look like this:

```
Filter on Red as Measure = CALCULATE(
                          SUM('Fact Sale'[Quantity]),
                          'Dimension Stock Item'[Color] = "Red"
                          )
```

Apart from the name and type of calculation, the expression used for both is identical. However, the calculations do not behave the same and that is due to different context.

Both calculations are created on the 'Fact Sale' table and specify a filter condition on a column in a different table. This works because you have defined a relationship between the 'Fact Sale' table and 'Dimension Stock Item' table in the data model.

Both calculations use the CALCULATE function to invoke the filter function as one of the easiest ways to apply a filter.

Both calculations are added to a pivot table (Figure 3-6) along with a nonfiltered implicit measure showing the sum of the quantity column.

Color	Quantity	Filter on Red as Column	Filter on Red as Measure
N/A	5,745,694		29,033
Black	1,101,101		29,033
White	873,153		29,033
Blue	605,738		29,033
Gray	309,847		29,033
Light Brown	274,500		29,033
Red	29,033	29,033	29,033
Yellow	11,562		29,033
Steel Gray			29,033
Total	**8,950,628**	**29,033**	**29,033**

Figure 3-6. *Pivot table showing output using [Filter on Red as Column] and [Filter on Red as Measure]*

The first column called Quantity is the [Quantity] field from the 'Fact Sale.' This is not one of our calculations; instead it's the field added to the pivot table to provide a baseline value as a reference for the two calculations.

Because this column uses a numeric datatype, a default summarization behavior of SUM is implicitly applied to these results, and the only context being applied to these results is query context from the row and column headers of the pivot table.

No DAX is required to generate this behavior. A column property of the data model allows you to define a default summarization for numeric columns. If no default is defined, the data model assumes that SUM should be the default behavior.

The next column in the pivot table is an implicit measure using the calculated column called [Filter on Red as Column], which only shows a value on the line where the row header is Red. The last column is the [Filter on Red as Measure] column, which shows the value of 29,033 repeated over and over, including on the last line, which is a Total row that is not filtered by a color.

Calculated Column

Let's look at what is happening with the [Filter on Red as Column] calculated column. In this case, we use the CALCULATE function. Let's use an explicit filter expression with this function to specify a rule that the 'Dimension Stock Item'[Color] must be "Red".

The SUM function doesn't understand row context, so the CALCULATE function automatically converts the row context into a filter context that allows the SUM function to access the data it needs to complete. See "Context Transition" at the end of this chapter, which explains this in more detail.

The calculation is evaluated at the point data is read into the data model, and for every row in the 'Fact Sales' table, the calculation is computed and the result is returned as the value for the new column. The explicit rule that 'Dimension Stock Item'[Color] must be "Red" is added to the filter context for each execution of the calculation.

The 'Fact Sales' table has 228,265 rows, so logically, this calculation is computed 228,265 times with each computation only having access to information from its current, or related, row.

Nothing else is added to, or removed from, the filter context, so if a row in the 'Fact Sales' table happens to be a red stock item, the calculation returns the SUM of the [Quantity] column, otherwise the calculation returns a blank.

The SUM function requires a column to be passed to it to produce a value. For each computation of all 228,265 rows, the SUM function uses a 'Fact Sale'[Quantity] column that is filtered by the filter context.

Once the calculated column has been added to the model, the data is visible in the Data View. This shows that only some rows have a value in the [Filter on Red as Column] column. These are rows that belong to stock items that are "Red". Most rows in this dataset will have blank values.

In Figure 3-7, two highlighted rows show values in the final column, whereas other rows show a blank cell. This is the only time the calculated column is computed and it explains why, when the column is added to the pivot table, the result shown in the table is only for data where the row header is "Red".

```
Filter on Red as Column = CALCULATE(
                               SUM('Fact Sale'[Quantity]),
                               'Dimension Stock Item'[Color]="Red"
                           )
```

Sale Key	Invoice Date Key	Quantity	Stock Item Key	Filter on Red as Column
437	Thursday, 3 January 2013	7	150	
449	Thursday, 3 January 2013	7	147	7
347	Thursday, 3 January 2013	7	88	
344	Thursday, 3 January 2013	7	199	
365	Thursday, 3 January 2013	8	160	
393	Thursday, 3 January 2013	8	154	
409	Thursday, 3 January 2013	8	195	
403	Thursday, 3 January 2013	8	152	8
354	Thursday, 3 January 2013	8	173	
401	Thursday, 3 January 2013	8	105	

Figure 3-7. *Sample of 'Fact Sale' table with new [Filter on Red as Column] calculated column*

Note   Setting a default summarization on a calculated column that uses the CALCULATE function has no effect. The expression used in the CALCULATE function determines the result for each cell, including totals.

The [Filter on Red as Column] calculated column only applies a very basic filter. An array of filter functions in DAX allows you to set more sophisticated filter behavior, which I cover in later chapters.

When you drag the calculated column to the pivot table, an implicit calculated measure is generated and used by the pivot table.

Calculated Measure

Let's look at the final column of the pivot table in Figure 3-8, which uses the [Filter on Red as Measure] calculated measure. This now returns values on rows with row headers for values other than "Red", and the very bottom row returns the same value as those higher in the column. You might expect the value shown in the final column to at least be the total of the values for that column.

Even though the formulas used for the calculated measure and calculated columns are identical, the difference is context. When the calculated column is computed, each individual calculation has row context and is logically computed 228,625 times.

Color	Quantity	Filter on Red as Column	Filter on Red as Measure
N/A	5,745,694		29,033
Black	1,101,101		29,033
White	873,153		29,033
Blue	605,738		29,033
Gray	309,847		29,033
Light Brown	274,500		29,033
Red	29,033	29,033	29,033
Yellow	11,562		29,033
Steel Gray			29,033
Total	**8,950,628**	**29,033**	**29,033**

Figure 3-8. *Pivot table showing output using [Filter on Red as Column] and [Filter on Red as Measure]*

The calculated measure is computed only ten times for this visual: once for each cell in the pivot table that has a row header (the nine different colors), and once for the total row at the bottom. When the calculated measures are evaluated, there is no row context. Each calculation starts with an empty filter context.

How does DAX arrive at the value of 29,033 for each execution of this calculated measure?

For the top nine rows, an *implicit* column filter-based rule is added to the empty filter context for each execution. This implicit filter rule comes from the first column in the table and states that the value in the 'Dimension Stock Item'[Color] column must match the value in the relevant row header.

However, an *explicit* column filter rule is also coded into the calculated measure using the same column. This replaces the implicit filter rule, meaning the filter context still has a single-column filter rule on this column but now uses the rule defined explicitly.

The calculation on the bottom row does not have an implicit column-based rule added to its filter context, but it still adds the explicit column filter to its filter context before it runs its SUM function. Therefore, it returns the same value as those above it.

Once the filter context is finalized, it contains a single-column filter rule over the 'Dimension Stock Item'[Color] column. The SUM function now returns 29,033 using the surviving values in the [Quantity] column. When this is repeated for every cell in the pivot table column, you get the same result, even for the grand total line.

This may seem odd, but hopefully it helps you understand how filter context is affecting the core calculation.

In case you are interested, Listing 3-5 shows the T-SQL equivalent of our calculated measure.

Listing 3-5. T-SQL Version of DAX Calculated Measure

```
SELECT
        SUM ( F.[Quantity] ) AS [Filter on Red as Measure]
FROM FACT.Sale AS F
        LEFT OUTER JOIN Dimension.[Stock Item] AS D
                ON F.[Stock Item Key]=D.[Stock Item Key]
WHERE
        D.[Color] = 'Red';
```

If you run this T-SQL query ten times (once for every cell), you get the same 29,033 result every time.

The T-SQL in Listing 3-5 uses a LEFT OUTER JOIN. This is also the join type that DAX would use in this particular case. Be aware that depending on the filters and DAX, sometimes this might be INNER JOIN.

Hardcoded Example

Another example of filter context is to hardcode both a calculated column and a calculated measure to return a value of 1.

Consider the following two calculations. The first is a *calculated column* while the second is a *calculated measure*:

```
Hardcoded Calculated Column = 1
```

```
Hardcoded Calculated Measure = 1
```

When we add these two calculations to the same pivot table we get the result shown in Figure 3-9.

Color	Quantity	Hardcoded Calculated Column ▼	Hardcoded Calculated Measure
N/A	5,745,694	95,810	1
Black	1,101,101	52,872	1
White	873,153	34,563	1
Blue	605,738	24,113	1
Gray	309,847	9,472	1
Red	29,033	5,224	1
Light Brown	274,500	4,114	1
Yellow	11,562	2,097	1
Steel Gray			1
Total	**8,950,628**	**228,265**	**1**

Figure 3-9. *Pivot table showing calculations using hardcoded values*

As expected, the calculated column executes 228,265 times and returns the value of 1 to every row. When added to the pivot table, an implicit calculated measure is generated using the default summarization behavior of SUM. The filter context for this implicit measure receives a column filter rule from the first column of the pivot table. This is a roundabout way of producing a ROWCOUNT measure.

The hardcoded calculated measure has ten cells to produce a value for, so logically, it computes the simple calculation ten times. The result as shown is the hardcoded value for 1 appearing in each cell—including the Total.

Row Context

Row context is effective when you create calculated columns or execute code inside an iterator function. Just as calculated measures perform a logical computation for each cell in a pivot table or report visual, calculated columns perform a logical computation to generate a value for every row in your calculated column. If you add a calculated column to a table with 250 rows, the calculation is executed 250 times. Each execution has a slightly different row context.

A key difference between calculated columns and calculated measures is the point in time the calculation takes place. Calculated columns are computed at the point they are created or modified and after data has been physically loaded or refreshed into the data model, whereas calculated measures recompute any time a filter is changed that might affect the value.

Row context allows the calculation to quickly access value from the same row in the current table or from values that might exist in rows in related tables via relationships.

A simple example of row context using the WideWorldImportersDW dataset (Figure 3-10) is to create a calculated column that returns a value based on multiplying the [Unit Price] and [Quantity] columns together.

| | | Total Price = 'Fact Sale'[Unit Price] * 'Fact Sale'[Quantity] |

Sale Key	Invoice Date Key	Quantity	Unit Price	Total Price
228265	Tuesday, 31 May 2016	8	32	256
228264	Tuesday, 31 May 2016	40	102	4080
228263	Tuesday, 31 May 2016	9	35	315
228262	Tuesday, 31 May 2016	1	25	25
228261	Tuesday, 31 May 2016	6	25	150
228260	Tuesday, 31 May 2016	24	18	432
228259	Tuesday, 31 May 2016	3	13	39

Figure 3-10. *Sample of 'Fact Sale' with new calculated column*

Figure 3-10 shows a new calculated column called [Total Price] that uses the [Unit Price] and [Quantity] columns from the same row in its calculation to produce each result.

The top row of in Figure 3-10 generates a value of 256 in the [Total Price] column, which is the result of 8 * 32.

Once the data has been loaded or refreshed, the value cannot be changed. If you have lots of calculated columns and many rows, your data load/refreshing takes longer. Another impact of adding calculated columns to the model is that additional memory is needed to store the computed results. Any additional calculation created that uses data from a calculated column treats the values as if they have been provided by the external data source.

If you have relationships defined in your data model, you can use the RELATED function to access columns from other tables to use in formulas in your calculated column.

An example of this is adding a calculated column to the 'Fact Sale' table that estimates the weight of each row based on the quantity information from the 'Fact Sale' table and combines this with the [Typical Weight Per Unit] column from the 'Dimension Stock Item' table. The formula for the calculated column uses data from two tables to produce the result.

Figure 3-11 highlights the relationship between the [Stock Item Key] columns from each table.

When the following calculated column is added to the 'Fact Sale' table, it demonstrates row context:

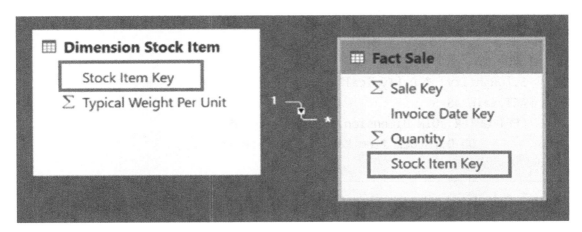

Figure 3-11. *Relationship View between 'Dimension Stock Item' and 'Fact Sale'*

```
Estimated Weight =
        'Fact Sale'[Quantity] *
                RELATED(
                    'Dimension Stock Item'[Typical Weight Per Unit]
                    )
```

You can see the new calculated column in the Data View in Figure 3-12.

Sale Key	Invoice Date Key	Quantity	Stock Item Key	Estimated Weight
1	Tuesday, 1 January 2013	10	153	150
2	Tuesday, 1 January 2013	9	170	1.35
3	Tuesday, 1 January 2013	9	210	0.45
4	Tuesday, 1 January 2013	3	106	1.05
5	Tuesday, 1 January 2013	96	14	9.6
6	Tuesday, 1 January 2013	5	90	1.75
7	Tuesday, 1 January 2013	2	170	0.3

Figure 3-12. *Sample output of 'Fact Sale' with the new calculated column*

IntelliSense only suggests columns in the 'Dimension Stock Item' table once you use the RELATED function. Relationships in the data model help control objects offered by IntelliSense.

The T-SQL equivalent of this calculation is issuing the following command for each row, although the S.[Stock Item Key] would be hardcoded to the value from the column defined in the relationship:

```
SELECT TOP 1
     S.[Quantity] * D.[Typical Weight Per Unit] AS [Estimated Weight]
FROM FACT.Sale AS S
     LEFT OUTER JOIN Dimension.[Stock Item] AS D
          ON D.[Stock Item Key] = S.[Stock Item Key]
```

Note   Calculated columns cannot access data from the row above or below the current row using a current row +/-1 instruction.

Although this note is not entirely true, it's not as straightforward as what you might encounter in other languages. You can access data from other rows in your table in calculations that use explicit DAX functions, such as FILTER, which I cover in Chapter 6 as a more advanced scenario.

Assume there is no concept of ordering when loading tables, so the previous/next row cannot be guaranteed. As soon as a row has been read and broken apart to populate the various column indexes during the data load process, it is discarded.

It is possible to logically reconstruct rows when you are required to, but this can be computationally expensive, especially if the table has many columns.

Iterators

Several DAX functions rebuild rows based on information stored in the column indexes. These are known as *iterators*, or *X functions*, because these functions typically have an X as the last character of the function name. Here are some of the iterator functions:

- AVERAGEX: Calculates the average (arithmetic mean) of a set of expressions evaluated over a table.

- COUNTAX: Counts the number of values that result from evaluating an expression for each row of a table.

- COUNTX: Counts the number of values that result from evaluating an expression for each row of a table.

- GEOMEANX: Returns the geometric mean of an expression value in a table.

- MAXX: Returns the largest numeric value that results from evaluating an expression for each row of a table.

- MEDIANX: Returns the 50th percentile of an expression value in a table.

- MINX: Returns the smallest numeric value that results from evaluating an expression for each row of a table.

- PRODUCTX: Returns the product of an expression value in a table.

- RANKX: Returns the rank of an expression that is evaluated in the current context in the list of values for each row in the specified table.

- SUMX: Returns the sum of an expression evaluated for each row in a table.

- FILTER: Returns a table that has been filtered.

Iterators are functions that process once for every row over any table passed to the function. A process can be a calculation, or in the case of FILTER, it applies a set of Boolean rules to filter the table it uses to return a filtered table.

Iterators can be nested many times. Each instance of an iterator keeps its own set of row context. Expressions in a nested iterator can access values from a row context from an outer iterator. These are often described as being the equivalent of the inner and outer loops you might see in other languages:

```
foreach (row outer in TableA)
{
        foreach (row inner in TableB)
        {
                return expression;
        }
}
```

Iterators allow you to solve sophisticated data problems that can't easily be solved using functions like SUM, COUNT, or AVERAGE, but they can cause an extra load on CPU processing.

Note   SUM/COUNT/AVERAGE are just syntax sugar for SUMX/COUNTX/ AVERAGEX. In the case of a single column, there is no performance advantage to using SUM('Table'[Column]) instead of SUMX('Table', [Column]).

How Data Is Stored in DAX

Another way to think about context in DAX is to consider how the underlying engine stores data once it has imported it. Looking at context in this way may help you appreciate the order and effect different types of context have on your calculations.

Your source data is typically stored in rows or batches of rows. Each row may have many columns, fields, or properties. When you import data to a DAX model, the data source is read into the data model row by row.

Table 3-4 is an example dataset based on US population estimates from the US Census Bureau. The raw dataset used here is only nine rows and has not been aggregated.

Table 3-4. *US Population Estimates*

US State	Age Band	2015 Population	2016 Population
California	Less than 25	25,947,866	25,748,276
California	Less than 50	26,922,326	27,188,676
California	50 or Over	24,805,956	25,245,954
Florida	Less than 25	11,676,422	11,729,046
Florida	Less than 50	12,627,122	12,853,344
Florida	50 or Over	16,049,448	16,501,676
Texas	Less than 25	19,885,656	20,033,802
Texas	Less than 50	18,700,892	19,046,066
Texas	50 or Over	16,036,504	16,412,550

When this dataset is imported into DAX, the import engine reads and processes the data row by row. A calculated column added to this dataset determines what the differences between the 2015 and 2016 population might look like:

```
Population Difference =
'Population by Age'[2016 Population] - 'Population by Age'[2015 Population]
```

When the first row is imported, the calculation can be computed easily because the engine has access to every value in the row, including the [2016 Population] and [2015 Population] columns used in the formula. The single-value result of the calculation is then stored in a new column on this row.

The calculation does not use any filter statements, so the filter context is empty. Each row is read in isolation, so you have no access to any information in any other rows (except through data model relationships with other tables).

Column Indexes

The next step of the import process is to break apart each of the columns into its own data structure. DAX keeps and maintains a separate storage structure for every value in the [US State] column, maintains a separate data structure for the [Age Band] column, and so on. I call these data structures column indexes. The data model applies several types of compression algorithms to each of the column indexes, which I will not detail here, but the key point is that each column index retains enough information to allow it to understand the original row number each value belonged to.

As I mentioned, the column indexes contain a reference to original row numbers, so although it is possible for the engine to stitch rows back together at runtime, it becomes an expensive operation in terms of performance. The DAX engine can compute most calculations very quickly using just a few column indexes without needing to reconstruct large numbers of complete rows.

Let's look at an example of how DAX uses the column indexes to generate a result. Say you would like to create a pivot table that uses the nine rows of population estimates and focus on the [2015 Population] broken down by [Age Band]. A pivot table would look like Figure 3-13. The top value of 57,509,944 has been highlighted; it is the combined total of three underlying rows that originally belonged to California, Florida, and Texas.

Age Band ▲	2015 Population
Less than 25	57,509,944
Less than 50	58,250,340
50 or Over	56,891,908
Total	**172,652,192**

Figure 3-13. *Pivot table using the [2015 Population] calculated measure*

The formula for the calculated measure is simple enough. This is the DAX to create this as an explicit measure:

```
2015 Population = SUM('Population by Age'[2015 Population])
```

The SUM function is passed a reference to the [2015 Population] column. The filter context in effect is that SUM should only consider rows where the [Age Band] has a value of "Less than 25".

Before the SUM function can begin, the engine first needs to scan the [Age Band] column index to obtain the original row numbers for all values it has for "Less than 25". These are likely to be sorted and grouped together, so the scan will quickly return a list of row numbers, which in this case are rows 1, 4, and 7 (Table 3-5). The next step is to allow the SUM function to proceed, but only for values that have these three numbers as row numbers.

Table 3-5. *Rows Used by SUM Function Filtered by [Age Band]*

Row Number	Age Band	Row Number	2015 Population
1	Less than 25	1	25,947,866
4	Less than 25	4	11,676,422
7	Less than 25	7	19,885,656

Once you have the [2015 Population] for rows 1, 4, and 7, the engine can return a value of 57,509,944 to the appropriate cell (Figure 3-14).

This is repeated for the second row, but in this case the row numbers retrieved from the [Age Band] column index for "Less than 50" are rows 2, 5, and 8.

The calculation then identifies that rows 3, 6, and 9 are required for the "50 or Over" values.

For the final row, there is no need to scan the [Age Band] column index, so all values from the [2015 Population] column are passed to the SUM function for the output of 172,652,192.

Age Band	2015 Population
Less than 25	57,509,944
Less than 50	58,250,340
50 or Over	56,891,908
Total	**172,652,192**

Figure 3-14. *Pivot table using the [2015 Population] calculated measure*

In summary, to compute the numbers for all but the total row, the engine needed to access two column indexes and no more. It did not need to scan or read the column indexes for [US State] or [2016 Population]. The value returned for the final row could be determined by reading just the [2015 Population]. The only context that was in effect was query context.

The underlying query engine may approach these steps in a way that is more efficient, but if you think about what is logically required by the engine to produce the final value for each cell, it may help to understand the effect of the three types of context.

Context Transition

On occasion, you need to convert a row context to a filter context. This is because some DAX functions do not understand row context.

The row context example from earlier in the chapter used the following formula:

```
Estimated Weight =
    'Fact Sale'[Quantity] *
            RELATED(
                'Dimension Stock Item'[Typical Weight Per Unit]
                )
```

This calculation doesn't use DAX functions such as SUM or AVERAGE, so it can compute a result that is meaningful and expected.

Let's look at what happens when you use a DAX function that does not understand row context by adding the following two calculated columns to the 'Fact Sale' table.

```
New Quantity Column 1 = SUM ( 'Fact Sale'[Quantity] )
```

```
New Quantity Column 2 = CALCULATE( SUM( 'Fact Sale'[Quantity] ) )
```

Both calculated columns use the SUM function and the same column reference is used for both. The only difference is the second calculated column is using the CALCULATE function around the SUM function. The results are shown in the two right-hand columns in Figure 3-15.

Sale Key	Invoice Date Key	Quantity	New Quantity Column 1	New Quantity Column 2
1	Tuesday, 1 January 2013	10	8,950,628	10
2	Tuesday, 1 January 2013	9	8,950,628	9
3	Tuesday, 1 January 2013	9	8,950,628	9
4	Tuesday, 1 January 2013	3	8,950,628	3
5	Tuesday, 1 January 2013	96	8,950,628	96
6	Tuesday, 1 January 2013	5	8,950,628	5
7	Tuesday, 1 January 2013	2	8,950,628	2
8	Tuesday, 1 January 2013	4	8,950,628	4

Figure 3-15. Sample of 'Fact Sale' showing new calculated columns

For the [New Quantity Column 1] column, you can see the value 8,950,628 repeated in every row of the table for this column. The filter context is empty for each execution of this version of the calculation, so the SUM function has access to every row in the 'Fact Sale'[Quantity] column.

The [New Quantity Column 2] column produces a different result that highlights the effect of context transition. The CALCULATE function knows it is being used in a calculated column, so it converts the existing row context for each line into a column-based filter rule that is then added to the filter context. This time around, each execution of the SUM function runs using a heavily filtered version of the 'Fact Sale'[Quantity] column.

Context transition can only convert row context to filter context. Context is never converted the other way.

CHAPTER 4

Summarizing and Aggregating

When it comes to data modeling, a common requirement is to generate summary versions of existing tables. This chapter explores some of the options available to you in DAX.

There are plenty of reasons why you might consider adding a summary table to your data model. Used well, summary tables can greatly improve performance. A large table made up of millions of transactions with thousands of rows per day can be a great candidate for a summary table that may only have one row per day. Metrics can be calculated while aggregating and stored as columns to provide most of the same reporting that might otherwise take place over the raw data. Visuals that use such a summary table are much faster as long as you only need to plot by day, month, or year.

It's important to be aware that when you create a summary table like this, you lose the ability to cross filter on any level of aggregation not included in the summary table. But with some thought and planning, you can optimize a Power BI report to provide a much better experience for your end users.

Other uses of a summary table may revolve around helping you understand some behavioral features in your dataset; for instance, when was the first and/or last time a customer made a purchase using the MIN and MAX functions over a purchase date column? You can join the resultant summary table back to the original table as a smart way of filtering records dynamically based on the first/last purchase date, which may be different for each customer. You can also use a summary table to find the top ten products based on sales using grouping and ranking together.

These are the main DAX functions that provide the ability to create summary tables:

- SUMMARIZE

- SUMMARIZECOLUMNS

- GROUPBY

© Philip Seamark 2018
P. Seamark, *Beginning DAX with Power BI*, https://doi.org/10.1007/978-1-4842-3477-8_4

This chapter looks at these functions in more detail and covers the differences between each with suggestions as to when you might choose one over the other. For simple examples over small sets of data, they are practically interchangeable. However, for more complex scenarios using larger datasets, choosing the right function can make a significant difference.

The summarization functions all return a new table as their output. They are often used as the main function in part of a calculate table expression as well as part of complex working logic inside a calculated measure.

The functions are similar in that you can specify columns to be used to group by. If no columns are specified, the function only returns a single-row table. If one column is used to group by, the number of rows reflect the total number of unique values in that column. The number of rows may grow as additional columns are added, but not always. Adding a calendar month column to a summarization function that is already grouping by calendar day does not add any rows to the final output.

The columns used to group can exist across multiple tables, as long as you have defined relationships appropriately. These are typically "upstream" tables that can also be referred to as tables on the "one side" of a DAX relationship.

I use the same example to demonstrate each of the summarization functions. These all use the WideWorldImportersDW dataset and create a summary over the 228,265-row 'Fact Sale' table and group by the [State Province] and [Calendar Month] columns. An aggregation column that carries a sum over the 'Fact Sale'[Quantity] column is added to each summary table.

Listing 4-1 shows what the T-SQL-equivalent statement for this looks like:

Listing 4-1. T-SQL Baseline Example

```
SELECT
     [Date].[Calendar Month Label],
     [City].[State Province],
     SUM(Sale.Quantity) AS [Sum of Qty]
FROM [Fact].[Sale] AS [Sale]

     INNER JOIN [Dimension].[Date] AS [Date]
          ON [Date].[Date] = Sale.[Delivery Date Key]

     INNER JOIN [Dimension].[City] AS [City]
          ON [City].[City Key] = Sale.[City Key]
```

```
GROUP BY
    [Date].[Calendar Month Label],
    [City].[State Province]
```

The resulting table contains 2,008 rows, down from nearly quarter of a million rows, which is a ratio of 111:1, or is a new table that is less than 1% the size of the original.

Any visual using this summary table has less work to do to generate values when used in a report. If the report only ever filters by the [Calendar Month Label] or [State Province] fields, then this is a good thing. However, if you need to represent other levels of filtering, then consider modifying the summary table or creating additional summary tables.

The SUMMARIZE Function

The first function we look at is the SUMMARIZE function. The syntax for calling this function is as follows:

```
SUMMARIZE (<table>, <groupBy_columnName1>..., <name1>, <expression1> ...)
```

The first parameter required is a value for <table>. This can be the name of any table that exists in your model, which can include other calculated tables. It can also be a function that returns a table, or a table expression stored in a variable. You can only provide one table.

Note   You cannot use aggregation functions in a SUMMARIZE function over the output of another SUMMARIZE function in the same statement. If you need to perform a multilayer aggregation, you need to use the GROUPBY function, otherwise create two table statements.

The second parameter can be the name of any column from the table specified as the first parameter or a column from any table with a relationship defined in your data model. You can continue to pass column references as the third, fourth, and fifth parameters (and so on). The DAX engine is smart enough to be able to figure out what to do based on the type of value you are passing to the function. As long as you continue to pass table columns, the function knows that these are the columns you wish to group by.

You are not required to pass any column references to the function if you don't want to. If your intention is to simply group your base table to a single line, you can start to provide <name> and <expression> pairs immediately after the first argument.

Once you stop passing table columns, you can start adding pairs of parameters to the function. These pairs exist in the form of a text value and a DAX expression. The <name> value is text that you use to set the name of the column. You need to encapsulate the text in double quotes. The other parameter in the pair is the <expression> value, which is a valid DAX calculation. This is the aggregation component of the summary table. This example uses a single pair of <name>/<expression> parameters, but like the <groupBy_columnName> arguments, it's possible to pass additional parameters as long as they match the <name>/<expression> signature.

The example in Listing 4-2 has been formatted in a way that makes it easier to follow.

Listing 4-2. DAX Example Using the SUMMARIZE Function

```
Summary Table using SUMMARIZE =
    SUMMARIZE(
        -- Table to Summarize --
        'Fact Sale',
        -- Columns to Group by --
        'Dimension Date'[Calendar Month Label],
        'Dimension City'[State Province],
        -- Aggregation Columns --
        "Sum of Qty",SUM('Fact Sale'[Quantity])
        )
```

Let's map the parameters in this example back to the syntax specification so you can see how the function works (Table 4-1).

Table 4-1. *Syntax Mapping of the SUMMARIZE Function*

Syntax	Example	
<table>	'Fact Sale'	
<groupBy_columnName1>	'Dimension Date'[Calendar Month Label]	
<groupBy_columnName2>	'Dimension City'[State Province]	
<name1>, <expression1>	"Sum of Qty"	SUM('Fact Sale'[Quantity])

The SUMMARIZE function returns a new table that otherwise has the appearance of any other table in your data model. You can use this table for visuals and filters and can relate it to other tables. You can add calculated columns and calculated measures. You can also now create calculated tables using the table created by the output of a SUMMARIZE function as your base table if you want to create a second layer of aggregation over your additional data.

Figure 4-1 has a blank value in the top-left row of the results. This means that there are records in the 'Fact Sale' table that have no value in the [Delivery Date Key] column or there are values in this column that don't exist in the 'Date Dimension'[Date] column. These are kept in the result set, and in this figure you can see the SUM of the original Quantity column add to 790.

Calendar Month Label	State Province	Sum of Qty
	New York	790
CY2013-Jan	New York	9237
CY2013-Feb	New York	8804
CY2013-Mar	New York	11780
CY2013-Apr	New York	9314
CY2013-May	New York	11570
CY2013-Jun	New York	10205
CY2013-Jul	New York	14740

Figure 4-1. *Sample output using the SUMMARIZE function*

One way to better understand the rows involved that have no date is by adding an additional aggregation column to the summary table using the <name>/<expression> pair like this:

```
"Count of Rows", COUNTROWS('Fact Sale')
```

This adds a fourth column to the calculated table that carries a value that shows the number of rows for each grouping, potentially providing extra insight as to many records may be affected by the mismatch.

Alternately, you can use the FILTER function to generate a calculated table showing every value in the 'Fact Sale'[Invoice Date Key] column that doesn't have a corresponding value in the 'Dimension Date'[Date] column. The code for this might look like

```
Table of Missing Data =
    FILTER(
        'Fact Sale',
        ISBLANK(
            RELATED('Dimension Date'[Date])
                )
            )
```

This generates a new calculated table in the model that is a filtered copy of 'Fact Sale' and only has rows where there are no matching rows in the 'Dimension Date' table.

The results of this useful debugging technique are 284 records in 'Fact Sale' that have a blank value in the [Delivery Date Key] column. These probably represent recent sales that have yet to be delivered.

The provides a DAX equivalent to a T-SQL LEFT JOIN query that is used to identify the values/rows from the left table that have no matching values/rows from the right query.

This calculated table that has been created can be removed once the data issue investigation work is complete.

Relationships

This example does not provide the SUMMARIZE function with any information on how to join the 'Fact Sale' table to the 'Dimension Date' and 'Dimension City' tables, yet the function is still able to group the data accurately the way we want using columns spanning three tables. In T-SQL statements, we provide hints on how to join tables using the ON and WHERE clauses, whereas the DAX engine uses the Relationship mechanism to determine how to connect the tables at query time.

In this case, the data model has an existing relationship defined between the 'Fact Sales' table and the 'Dimension Date' table. This relationship specifies that the [Delivery Date Key] column from the 'Fact Sale' table relates to the [Date] column from the 'Dimension Date' table.

The other relationship in play for this calculation is between the 'Fact Sale' table and the 'Dimension City' table using the [City Key] column on both sides. The SUMMARIZE function automatically uses the active relationships to construct the query required.

In DAX, it's possible to define multiple relationships between the same two tables. Only one can be active at any time, and the decision as to which relationship is active needs to be manually set by the author of the data model. It is the active relationship that is used by default that, in this case, joins the 'Fact Sale' table to the 'Dimension Date' table using the [Invoice Date Key] column.

Alternative Relationships

If you want to create a summary table using the same columns but you want to aggregate 'Fact Sale' data using a using a different date column, such as [Invoice Date Key], you can either manually edit the existing relationships between 'Fact Sales' and 'Dimension Date' to specify the appropriate relationship to be the active relationship or you can write your calculated table formula a different way to take advantage of an alternative relationship. Remember that in DAX, the relationship that is marked as 'Active' is the one used by default when you are writing formulas that involve multiple tables.

Listing 4-3 shows what the calculated table looks like if you want to use one of the inactive relationships. Your data model has an inactive relationship between 'Fact Sale' [Invoice Date Key] and 'Dimension Date'[Date].

Listing 4-3. SUMMARIZE Using the USERELATIONSHIP Function

```
Summary Table using SUMMARIZE =
    CALCULATETABLE(
        SUMMARIZE(
            -- Table to Summarize
            'Fact Sale',
            -- Columns to Group by
            'Dimension Date'[Calendar Month Label],
            'Dimension City'[State Province],
            -- Aggregation Columns
            "Sum of Qty",SUM('Fact Sale'[Quantity])
            ),
        USERELATIONSHIP(
            'Fact Sale'[Invoice Date Key],
            'Dimension Date'[Date])
        )
```

Here you have encapsulated your original DAX expression inside a CALCULATETABLE function. This allows a filter function to be used. The first parameter is the same as in the calculation you used earlier. The second parameter of the CALCULATETABLE function is the USERELATIONSHIP filter function that instructs the DAX engine to connect 'Fact Sale' to 'Dimension Date' using the [Invoice Date Key] column.

The result of this query shows the same three columns, but this time the [Sum of Qty] column has results that represent the date of the invoice rather than the date of delivery. This approach does rely on a relationship to exist, even if it is inactive. If there is no relationship between 'Fact Sale' and 'Dimension Date' via [Invoice Date Key], the calculated table is not loaded to the table.

It might be useful to incorporate this into the name of the new calculated table in some way to make it easier for end users to understand how the data has been aggregated.

SUMMARIZE with a Filter

In Listing 4-4, let's extend this example by incorporating a filter into the logic of the calculated table.

The addition of a filter restricts the underlying data used for the summary table to a specific sales territory. Another enhancement adds a column to the output that performs a basic calculation to represent an average quantity.

Listing 4-4. SUMMARIZE Function Using a Filter

```
Summary Table using SUMMARIZE (Southwest) =
      SUMMARIZE(
          -- Table to Summarize
          FILTER('Fact Sale',RELATED('Dimension City'[Sales Territory]) =
          "Southwest"),
          -- Columns to Group by
          'Dimension Date'[Calendar Month Label],
          'Dimension City'[State Province],
          -- Aggregation Columns
          "Sum of Qty",SUM('Fact Sale'[Quantity]),
          "Average Qty", DIVIDE(
                            SUM('Fact Sale'[Quantity]),
                            COUNTROWS('Fact Sale')
                            )
          )
```

To incorporate a filter into the calculation, the 'Fact Sale' table used in the first parameter is wrapped in a FILTER function. The FILTER function uses the RELATED function to apply a hardcoded filter to the 'Dimension City'[Sales Territory] column. The other enhancement is the additional <name>/<expression> to the aggregation columns that shows how you can use multiple aggregation functions (SUM and COUNTROWS) within the expression to achieve a slightly more complex result.

Another way to incorporate an explicit filter predicate is to use the CALCULATETABLE function as shown in Listing 4-5.

Listing 4-5. SUMMARIZE Using CALCULATETABLE to Filter

```
Summary Table using SUMMARIZE (Southwest) =
    CALCULATETABLE(
        SUMMARIZE(
            -- Table to Summarize
            'Fact Sale',
            -- Columns to Group by --
            'Dimension Date'[Calendar Month Label],
            'Dimension City'[State Province],
            -- Aggregation Columns --
            "Sum of Qty",SUM('Fact Sale'[Quantity]),
            "Average Qty", DIVIDE(
                                        SUM('Fact
                                        Sale'[Quantity]),
                                        COUNTROWS('Fact Sale')
                                        )
        ),
    'Dimension City'[Sales Territory] = "Southwest"
    )
```

To incorporate a filter into this calculation, the CALCULATETABLE function is used. The first argument is the SUMMARIZE function we used in Listing 4-4. The second argument passed to the CALCULATETABLE function is a basic filter statement. This filter means the calculation only uses data that matches this predicate; this is much like including a similar statement in the WHERE clause of a T-SQL query.

The SUMMARIZECOLUMNS Function

Another function you can use to aggregate tables in DAX is the SUMMARIZECOLUMNS function. This newer function has a slightly different signature in terms of syntax. The syntax is as follows:

```
SUMMARIZECOLUMNS(
    <groupBy_columnName1>...,
    <filterTable>...,
    <name1>,<expression1> ...
    )
```

Just as with the SUMMARIZE function, you can pass a dynamic number of parameters to this function. If the first parameter is a reference to a column, the SUMMARIZECOLUMNS function understands that this is a column you would like to group by. If you continue to pass column references, the function treats these as additional group by instructions.

Once you pass a FILTER function (or string value), the function understands that these references are instructions to either apply a filter to the aggregation or to start adding columns that will be the result of a DAX expression such as SUM, COUNT, MIN, and so on.

Using this syntax, the DAX you would use create the scenario using SUMMARIZECOLUMNS is shown in Listing 4-6.

Listing 4-6. DAX Example Using SUMMARIZECOLUMNS

```
Summary Table using SUMMARIZECOLUMNS =
    SUMMARIZECOLUMNS(
        -- Columns to Group by
        'Dimension Date'[Calendar Month Label],
        'Dimension City'[State Province],
        -- Aggregation Columns
        "Sum of Qty",SUM('Fact Sale'[Quantity])
        )
```

The difference between the SUMMARIZE example in Listing 4-4 and the SUMMARIZECOLUMNS function in this example is the absence of the first parameter in the SUMMARIZE function. This function does not need to pass a base table; however, the results are the same. The function can also group using columns from different, but related, tables. The DAX engine can use the active relationships between the 'Fact Sales' table and the two-dimensional tables to return the results required.

SUMMARIZECOLUMNS with a Filter

Another difference between SUMMARIZE and SUMMARIZECOLUMNS is in the latter, you can incorporate a filter as one of the parameters. Using the same requirement as earlier, Listing 4-7 summarizes and aggregates, but only data where the [Sales Territory] is "SouthWest".

Listing 4-7. The SUMMARIZECOLUMNS Function Using a Filter

```
Summary Table using SUMMARIZECOLUMNS (Southwest) =
    SUMMARIZECOLUMNS(
        -- Columns to Group by --
        'Dimension Date'[Calendar Month Label],
        'Dimension City'[State Province],
    -- Filter condition --
    FILTER(
        ALL(
            'Dimension City'[Sales Territory]),
            'Dimension City'[Sales Territory] = "Southwest"
            ),
        -- Aggregation Columns
        "Sum of Qty",SUM('Fact Sale'[Quantity])
        )
```

The calculation produces the same 164-row result as the example used with the SUMMARIZE function (Listing 4-4); however, there is no need to use the CALCULATETABLE function as a means of incorporating a filter. You can still wrap the SUMMARIZE function inside a CALCULATETABLE function if you prefer, and if you are using the data provided in the WideWorldImportersDW dataset, there is very little to separate the various approaches in terms of query runtime.

Under the cover, the SUMMARIZETABLE function appears to be doing fewer steps than the SUMMARIZE function; we will explore this type of analysis in more detail in Chapter 8.

The GROUP BY Function

The final aggregation function to look at is the GROUPBY function. This is its syntax:

```
GROUPBY(<table>, <groupBy_columnName1>..., <name1>,<expression1> ...)
```

GROUPBY is like the SUMMARIZE function in terms of syntax, and again, once you provide a single table reference as the first argument, you can pass any number of column references to let the GROUPBY function know the columns define the level of summarization.

Where GROUPBY differs from SUMMARIZE and SUMMARIZECOLUMNS is in the code passed to the function as part of any of the <expression> arguments. GROUPBY only works with the DAX iterator functions—so it uses SUMX rather than SUM and AVERAGEX rather than AVERAGE. This makes the GROUPBY function useful in specific scenarios.

If you wanted a calculated table to use the GROUPBY function to produce the same output as the SUMMARIZE and SUMMARIZECOLUMNS examples, you could write it as shown in Listing 4-8.

Listing 4-8. DAX Example Using the GROUPBY Function

```
Summary Table using GROUPBY =
    GROUPBY(
        -- Table to Group --
        'Fact Sale',
        -- Columns to Group By --
        'Dimension Date'[Calendar Month Label],
        'Dimension City'[State Province],
        -- Aggregation columns --
        "Sum of Qty",SUMX( CURRENTGROUP(),'Fact Sale'[Quantity] )
        )
```

The code is nearly identical to SUMMARIZE up to the final line, which adds the aggregation column. The DAX expression here is using the SUMX function instead of the SUM function. The SUMX function normally requires a table reference to be passed as the first argument, but here a CURRENTGROUP function is being passed instead. When the GROUPBY function was introduced to DAX, it came with a helper function called CURRENTGROUP(), which can be used in place of the original table.

The CURRENTGROUP() Function

The DAX iterator functions generally require a table reference to be passed as the first parameter. However, when used inside a GROUPBY function as is done here, the SUMX function must be passed to the CURRENTGROUP() function; any other value produces an error.

Think of the placement of the CURRENTGROUP() function as a mapping back to the table that was passed as the first parameter of the GROUPBY function. In this case, the table passed was 'Fact Sale', so the SUMX function, in effect, processes rows from the 'Fact Sale' table.

The earlier query at Listing 4-8 produces the same output as the previous examples using SUMMARIZE and SUMMARIZECOLUMNS; however, the results of running these three DAX functions side by side in a speed test are shown in Table 4-2.

Table 4-2. *Time in Milliseconds of Example Using Summary Functions*

DAX Function	Average Time (ms)
SUMMARIZE	40
SUMMARIZECOLUMNS	35
GROUP BY	130

These results are based on the same input and output, and with this very simple example, they show that SUMMARIZECOLUMNS is slightly quicker than SUMMARIZE, whereas GROUPBY is quite a bit slower.

This suggests that, for small and simple operations, the three functions are practically interchangeable, and you are unlikely to need to worry about which might be the best to use. However, you may prefer to use SUMMARIZE or SUMMARIZECOLUMNS simply for the slightly easier syntax and because you will not need to work with the iterator/CURRENTGROUP requirement.

You may be wondering what to use GROUPBY for given that it is likely to be slower and has a more complex syntax.

Two examples I focus on are its iterators and its ability to perform multiple groupings in the same operation.

GROUPBY Iterators

Because the GROUPBY function uses the DAX iterator functions, such as SUMX, AVERAGEX, MINX, and so on, it can unlock advantages these functions have over standard aggregation functions that allow you to perform calculations that involve other data from the same row.

An example that demonstrates this is extending the aggregation so it is a calculation involving multiple columns. This version multiplies the [Quantity] and [Unit Price] values together to produce a value for [Total Price], which can then be summed up.

The modified code is shown in Listing 4-9.

Listing 4-9. The SUMX Function Used Inside a GROUPBY Function

```
Summary Table using GROUPBY =
    GROUPBY(
        -- Table to Group --
        'Fact Sale',
        -- Columns to Group By --
        'Dimension Date'[Calendar Month Label],
        'Dimension City'[State Province],
        -- Aggregation columns --
        "Sum of Total Price",SUMX(
                        CURRENTGROUP(),
                        'Fact Sale'[Quantity] * 'Fact Sale'[Unit Price])
                        )
```

Without much change, the function now produces a result that allows you to sum multiple columns from each row. The only change has been to the last line of code. The <name> value has been changed to now show "Sum of Total Price". This has no impact on the calculation and is only good housekeeping. The main change is within the SUMX function; you can now write DAX expressions that include multiple columns from the same row—in this case, * 'Fact Sale'[Unit Price] is added to the <expression>. So, with a minor change to the calculation, you can unlock some interesting scenarios, and in this case, the change to the overall query time is undetectable.

SUMMARIZE and SUMMARIZECOLUMNS can also use the iterator functions to calculate over multiple columns from the same row. The equivalent calculation using SUMMARIZECOLUMNS is shown in Listing 4-10.

Listing 4-10. Example of Using the SUMX Function with SUMMARIZECOLUMNS

```
Summary Table using SUMMARIZECOLUMNS and SUMX =
    SUMMARIZECOLUMNS(
            -- Columns to Group By --
        'Dimension Date'[Calendar Month Label],
        'Dimension City'[State Province],
            -- Aggregation columns --
        "Sum of Total Price",SUMX(
                                'Fact Sale',
                                'Fact Sale'[Quantity] * 'Fact
                                Sale'[Unit Price])
                                )
```

Note   Be aware of the trap of thinking that SUMX (col1 * col2) is the same as SUM(col1) * SUM(col2). These produce quite different results. This is a common cause of confusion on internet help forums, particularly around sub/grand total lines.

This example can be extended to use columns from related tables in the same calculation. If the 'Dimension City' table has a column called [Weighting] that might carry a multiplier for each city, then as long as you have a valid and active relationship between the 'Fact Sale' and 'Dimension City' tables, you can extend the SUMX function to look more like this:

```
SUMX(
    CURRENTGROUP(),
    'Fact Sale'[Quantity] * 'Fact Sale'[Unit Price] *
    RELATED('Dimension City'[Weighting])
    )
```

Note the use of the RELATED function to allow the calculation to find and use values from the [Weighting] column as part of the calculation.

The GROUPBY Double Aggregation

Another feature of GROUPBY is the ability to perform multiple layers of aggregation in a single statement. For the first look at this, let's use the dataset in Table 4-3.

Table 4-3. *Simple Dataset to Help Demonstrate Double Grouping*

Category	Value
A	1
A	2
A	3
B	4
B	5
C	6
C	7

The challenge is to take the data in Table 4-3, find only the highest value for each category, and then add these together to produce a single output for the table. The result should be 3 + 5 + 7 = 15 because the max value for Category A is 3, the max value for Category B is 5, and the max value for Category C is 7. This involves two layers of aggregation. The first is to find a MAX value for a group, and the second is a SUM over the results of the MAX operation.

The SUMMARIZE, SUMMARIZECOLUMNS, and GROUPBY functions can all easily perform the first part of the requirement, which is to generate a three-row aggregation showing the maximum value for each of the categories.

It is the next step that trips the SUMMARIZE and SUMMARIZECOLUMNS functions up because it's not possible to layer or nest these functions in the same statement, whereas you can with GROUPBY.

The calculation using GROUPBY to meet this requirement is shown in Listing 4-11.

Listing 4-11. Calculated Table Using Two Layers of Summarization

```
Table =
GROUPBY (
    GROUPBY (
        'table1',
        Table1[Category],
        "My Max", MAXX ( CURRENTGROUP (), 'Table1'[Value] )
    )
    "Sum of Max values", SUMX ( CURRENTGROUP (), [My Max] )
)
```

There are two GROUPBY statements in this calculation. I refer to the first GROUPBY as the outer GROUPBY, whereas the second GROUPBY highlighted in Listing 4-11 is referred to as the inner GROUPBY.

In this example, the inner GROUPBY is computed first, which returns a table as output. The outer GROUPBY uses this output as the first parameter that is going to perform the second level of aggregation over the dataset.

The CURRENTGROUP() function appears twice in this calculation and this double grouping example provides a clearer example of the behavior of this function. Each instance of CURRENTGROUP() is paired with a relevant GROUPBY function. The CURRENTGROUP() in the highlighted section is used along with the MAXX iterator function as a reference to 'Table1', whereas the CURRENTGROUP() function in the final line is a reference to the output of the inner GROUPBY function.

The example could have used either the SUMMARIZE or SUMMARIZECOLUMNS function in place of the inner GROUPBY, however, only GROUPBY can be used as the outer level of summarization.

It would be better to use SUMMARIZE or SUMMARIZECOLUMNS as the inner layer of summarization in this example since you can meet this requirement without needing an iterator function.

Introducing variables to the computation also helps make the code more readable; one way you can write this is shown in Listing 4-12.

Listing 4-12. Improved Version of Calculated Table Using Two Layers of Summarization

```
Table =
    VAR myInnerGroup =
        SUMMARIZECOLUMNS (
            Table1[Category],
            "My Max",
            MAX ('table1'[Value])
            )
    RETURN
        GROUPBY(
            myInnerGroup,
            "Sum of max values",
            SUMX (
```

```
            CURRENTGROUP (), [My Max]
            )
        )
```

When you mix the power of iterator-based functions with the multilayer grouping ability of GROUPBY, you soon find it to be a useful summarization function for more complex scenarios despite the extra time it might take in some scenarios.

GROUPBY with a Filter

As with the earlier functions, let's look at how you might apply a filter to the GROUPBY function. Unlike with SUMMARIZECOLUMNS, there is no natural place in the parameter signature to pass a filter. Listing 4-13 shows the first of a pair of examples that restrict the standing example to where 'Dimension City'[Sales Territory] = "Southwest".

Listing 4-13. GROUPBY Function Using a Filter

```
GROUPBY(
    -- Table to Group --
    FILTER('Fact Sale',RELATED('Dimension City'[Sales Territory])="Southwest"),
    -- Columns to Group By --
    'Dimension Date'[Calendar Month Label],
    'Dimension City'[State Province],
    -- Aggregation columns --
    "Sum of Qty",SUMX(CURRENTGROUP(),'Fact Sale'[Quantity])
    )
```

Listing 4-13 uses the FILTER function as the first parameter of the GROUPBY function. The FILTER function returns a table, which is why this can still work in place of a specific table.

Listing 4-14 shows an alternative notation that uses the CALCULATETABLE function.

Listing 4-14. Example Using CALCULATETABLE to Filter a GROUPBY

```
CALCULATETABLE(
    GROUPBY(
        -- Table to Group --
            'Fact Sale',
            -- Columns to Group By --
```

```
        'Dimension Date'[Calendar Month Label],
        'Dimension City'[State Province],
        -- Aggregation columns --
        "Sum of Qty",SUMX(CURRENTGROUP(),'Fact Sale'[Quantity])
        ),
    FILTER('Dimension City',[Sales Territory]="Southwest")
    )
```

Here the filter is applied outside the GROUPBY function and is used as a parameter of CALCULATETABLE. This differs slightly from the first example in that you do not need to use the RELATED function for the cross-table filter to take effect. Both queries produce the same output and on the data volumes used in the WideWorldImportersDW, there was little to separate the two approaches in terms of performance, mostly due to the simple nature of the aggregation functions used.

Note   Both these examples use GROUPBY over a physical table. You should show preference to SUMMARIZE and SUMMARIZECOLUMNS when you are using physical tables. These examples are intended to show how you might apply a filter with the GROUPBY function if you need to.

Subtotal and Total Lines

A slightly more advanced use of these summarization functions is being able to inject additional rows into the output in the form of subtotals and grand totals. This is not generally a requirement when writing DAX calculations in Microsoft Power BI because Power BI takes advantage of these functions to generate totals when it is generating the dynamic DAX behind visuals.

If you are writing a DAX statement using the SUMMARIZE function using SQL Server Management Studio (SSMS), DAX Studio, SQL Server Reporting Services (SSRS), or other client tools, you can add the ROLLUP, ROLLUPGROUP, and ISSUBTOTAL functions to inject additional rows and information to the output. These queries work in Power BI, but you may need to filter them so as not to confuse the additional code that Power BI might add.

Let's use the same table from earlier, shown again here with a new column called Sub Category. This time you are going to group this data by Category and/or Sub Category and show a value that is a sum of the Value column. Additionally, you'll include subtotal and overall total lines as part of the output.

Table 4-4. *Simple Dataset to Help Demonstrate Double Grouping*

Category	Sub Category	Value
A	A1	1
A	A1	2
A	A2	3
B	B1	4
B	B2	5
C	C1	6
C	C2	7

This example uses the SUMMARIZE function to group the data using the Category column and includes an instruction to add an additional line to the output to carry a line for the overall total.

The first version of this query is as follows:

```
SUMMARIZE (
    table1,
    ROLLUP ( 'table1'[Category]),
    "Sum of Value", SUM ( 'table1'[Value] )
    )
```

The output from this calculation is shown in Table 4-5.

Table 4-5. *Output of the SUMMARIZE Function*

[Category]	[Sum of Value]
A	6
B	9
C	*13*
	28

The bottom row has no value in the Category column and shows the sum value of the other rows. This is the extra row added because the ROLLUP function was added to the calculation.

The value shown in the final line is influenced by the aggregation function used by the column, which in this case was a SUM. Adding two columns that use the MAX and AVERAGE aggregation functions helps show the behavior (Listing 4-15) and output (Table 4-6) of the ROLLUP row.

Listing 4-15. Using SUMMARIZE with ROLLUP

```
SUMMARIZE (
    table1,
    ROLLUP ( 'table1'[Category]),
    "Sum of Value", SUM ( 'table1'[Value] ),
    "Max of Value", MAX ( 'table1'[Value] ),
    "Average of Value", FIXED ( AVERAGE ( 'table1'[Value] ), 1 )
    )
```

Table 4-6. *The Output of SUMMARIZE with ROLLUP*

[Category]	[Sum of Value]	[Max of Value]	[Average of Value]
A	6	3	2.0
B	9	5	4.5
C	*13*	*7*	*6.5*
	28	*7*	*4.0*

The value of 7 in the bottom cell in the [Max of Value] column is not the result of a SUM function; rather, it inherits the default aggregation behavior used in the expression for the column. In this case, it shows the highest value from the underlying 'Table1'[Value] column.

The value in the bottom row of the [Average of Value] column shows that the ROLLUP function is running an aggregation over the raw data rather than an average of the three values shown above it. If it were performing an average of an average, the value would be 4.3.

The FIXED function around the AVERAGE function in the expression specifies that the output should be displayed with a specified number of decimal points. This example uses one decimal place.

The ISSUBTOTAL Function

You can add a column to the calculation to help identify which lines have been added using the ROLLUP function. Do this by adding the ISSUBTOTAL function to the expression of an aggregation column as follows:

```
SUMMARIZE (
    table1,
    ROLLUP ( 'table1'[Category]),
    "Sum of Value", SUM ( 'table1'[Value] ),
     "Is Category Subtotal", ISSUBTOTAL('table1'[Category])
    )
```

The output of this calculation is shown in Table 4-7.

Table 4-7. *Output of the SUMMARIZE Function Using ISSUBTOTAL*

[Category]	[Sum of Value]	[Is Category Subtotal]
A	6	False
B	9	False
C	*13*	*False*
	28	*True*

This calculation now includes a column with true/false values that help identify which rows have been added to the output because of the ROLLUP function. In this example, only the bottom row carries a value of true in this column as shown in Table 4-7.

The ISSUBTOTAL function becomes more useful when extra layers of grouping are introduced to a SUMMARIZE function using ROLLUP. Listing 4-16 uses the SUMMARIZE function to create a summary table (Table 4-8) that is grouped by two columns.

Listing 4-16. SUMMARIZE Using an Additional ROLLUP Parameter

```
SUMMARIZE (
    table1,
    ROLLUP ( 'table1'[Category], 'table1'[Sub Category] ),
    "Sum of Value", SUM ( 'table1'[Value] ),
    -----------
    "Is Cat SubTotal", ISSUBTOTAL('table1'[Category]),
    "Is Sub Cat SubTotal", ISSUBTOTAL('table1'[Sub Category])
)
```

Table 4-8. *Output of SUMMARIZE Using an Additional ROLLUP Parameter*

[Category]	[Sub Category]	[Sum of Value]	[Is Cat Sub Total]	[Is Sub Cat Subtotal]
A	A1	3	False	False
A	A2	3	False	False
B	*B1*	*4*	False	False
B	*B2*	*5*	False	False
C	*C1*	*6*	False	False
C	*C2*	*7*	False	False
A		*6*	False	*True*
B		*9*	False	*True*
C		*13*	False	*True*
		28	*True*	*True*

Here you have introduced an additional parameter to the ROLLUP function, which tells the SUMMARIZE function to now group by two columns rather than one. This adds a [Sub Category] column but also adds three additional lines of subtotal over the groupings of Category.

An additional aggregation column called [Is Sub Cat Subtotal] has been added using the ISSUBTOTAL function in the expression over the [Sub Category] column to show a true or false value. If the value in this column is false, the row is not a subtotal line of the [Sub Category] column. If the value is true, the line has been added for providing subtotal values.

Each subtotal line of [Category] has the following characteristics:

- The [Category] column is populated with a value.

- The [Sub Category] column is blank.

- The [Sum of Value] column carries the default aggregation result for the underlying rows for that category.

- The [Is Cat Subtotal] column is false.

- The [Is Sub Cat Subtotal] column is true.

The final line of the output happens to be the grand total line. It has the same characteristics as the subtotals over [Category] except

- The [Category] column is blank.

- The [Is Cat Subtotal] column is true.

This shows it is possible, with a combination of ROLLUP and ISSUBTOTAL, to enhance your output with additional information that can be useful in some scenarios. You can add additional functionality when using SUMMARIZE with the ROLLUPGROUP function, which when combined with ROLLUP, can be used to combine subtotals to achieve finer control over the final output.

The ROLLUP and ISSUBTOTAL functions are designed to work with the SUMMARIZE function only. The SUMMARIZECOLUMNS function uses a separate set of functions when you are adding lines to the output for showing totals. The main function to highlight is the ROLLUPADDISSUBTOTAL function that is designed to work with SUMMARIZECOLUMNS.

Listing 4-17 uses SUMMARIZECOLUMNS to yield the same output as in Listing 4-16.

Listing 4-17. Using SUMMARIZECOLUMNS with ROLLUPADDISSUBTOTAL

```
SUMMARIZECOLUMNS (
    ROLLUPADDISSUBTOTAL (
        Table1[Category], "Is Cat Sub Total",
        Table1[Sub Category], "Is Sub Cat SubTotal"
        ),
    "Sum of Value", SUM ( Table1[Value] )
    )
```

Note that the syntax is simpler than with SUMMARIZE even though you get the same number of rows and columns. In addition, the ROLLUPADDISSUBTOTAL can be used multiple times and combined with the ROLLUPGROUP function to provide rich control over subtotal and subtotal groupings.

CHAPTER 5

Joins

Joins in DAX

This chapter looks at options for using and combining data from different tables in calculations without necessarily using the relationship mechanism.

It's common for calculations to use data from multiple tables. Most scenarios can be covered automatically through the standard table relationship defined in the data model, but not in every case. If you need more flexibility, several DAX functions are available that support joining tables. Some of these provide an experience that should be familiar to you if you are used to writing T-SQL queries. I highlight the similarities and differences in this chapter.

Here are the functions that I cover:

- GENERATE and CROSSJOIN

- NATURALINNERJOIN

- NATURALLEFTOUTERJOIN

- UNION

- LOOKUPVALUE

Standard Relationship

If your data model contains more than one table, it's possible to define rule based links between tables using the standard DAX relationship mechanism. DAX relationships are a link between two tables that define rules on how the relationship should work. Calculations take advantage of these predefined relationships automatically, which can simplify how much code you need when you're writing most calculations.

There are some limitations in the standard DAX relationship mechanism that do not work for some scenarios.

© Philip Seamark 2018
P. Seamark, *Beginning DAX with Power BI*, https://doi.org/10.1007/978-1-4842-3477-8_5

Relationships must conform to the following rules:

- The relationship can only be one-to-one, or one-to-many.

- Only a single column from each table can be used.

- The match criteria must be an exact match equivalent of the = operator and cannot be based on other operators such as >, >=,<, or <= and so on.

- Self joins cannot be used. The two tables must be different.

The relationship can only be one-to-one, or one-to-many

This means that for at least one of the tables involved in the relationship, unique values must exist across every row in the table in the column used for the relationship. This is referred to as "the table on the one side of the relationship."

Values that are the same and exist in more than one row in the table are considered duplicates and generate an error. This is something to be mindful of for future data loads. A successfully created relationship confirms that the current dataset conforms. However, future data loads may introduce duplicated data and therefore generate an error during the refresh.

A common example of one-to-many relationships is between dimension and fact tables such as 'Dimension Date' and 'Fact Sale'. The date table contains a single row for every day, which carries a unique value in the [Date] column. This satisfies the requirement for the table on the one side of a relationship, whereas the 'Fact Sale' table can have many rows that carry the same date in the column used in the relationship.

This is a common pattern in data modeling. Fact (or event) tables often have relationships defined to dimension (or lookup) tables. The fact table can have many rows for the same date, customer, or location, with separate tables that contain a single row for every date, customer, or location. These tables typically sit on the one side of any relationship.

You can also consider the tables on the one side of the relationship as filter tables. These are ideal tables to use as the basis for slicers and filters in reports. Selections made on these slicers propagate via relationships to any related tables connected on the many side and filter fact data appropriately. This approach works well for analytics and reporting, and the pattern is often described as applying a star schema to your data. Relationships can be created between dimension tables to multiple fact tables. This allows filter selections to apply across large sections of data in the model.

Filters only apply in the direction of the one table to the many table. This means a slicer using a column in the 'Fact Sale' table does not automatically apply to calculations using data from the 'Dimension Date' table.

One-to-one relationships differ from one-to-many in that the column involved in each table must have unique values. This type of relationship effectively creates a single logical table, or in T-SQL, a full outer join across the two tables.

Only a single column from each table can be used

It's not possible to define a relationship using anything more than one column per table. If you have a scenario in your data in which you need to enforce a relationship that needs to use data from multiple columns, you can create a new column that concatenates these columns to a single value that you can then use in the relationship, as long as the new column still satisfies the rule of having unique values for any table being used on the one side.

When you're using this approach, be careful to avoid a key collision from two separate values combining to produce the same result. Combining values "AB" and "CDE" results in the same value as combining "ABC" and "DE". In this case, an unusual delimiter should reduce the risk of a key collision, for example, "AB" and "#" and "CDE" would now be different than "ABC" and "#" and "DE".

The match criteria must be an exact match

The default operator used between the tables involved in a relationship is the = operator. This means that for rows to be matched, the relationship only pairs rows from either table where there is an exact match in the values of the two columns involved. This can be a trap when you're working with similar-looking values.

It's possible to create a one-to-many (or one-to-one) relationship between a table that uses the Date datatype for the column on one side of the relationship and uses DateTime on the other. Only values at midnight in the column using the DateTime datatype can possibly match values from the column using the Date datatype. This can be confusing when you have two tables that appear to have data that should be matched and, although no errors appear, visuals show less data than you expect.

In this event, consider converting the column in the table that is using the DateTime datatype to use Date instead. If the time component is important, you can use a second column that carries just the time component to satisfy most reporting requirements.

Another common solution is to create a calculated column that only contains the date portion and then create a relationship using the calculated column.

This behavior works well for most scenarios, but occasionally you may want to create a match between tables using more than just the = operator. Sometimes using the >= or <= operators can be useful when you need to join rows that span a range of values. An example of this might be when, for each row in a table, you need to find rows in a separate table that are within a period of the first row, such as the prior 14 days.

Self joins cannot be used

You cannot create a relationship from a table connected back to itself. This type of relationship can be useful to represent hierarchies such as employee/manager scenarios in a table that contains employee data.

The employee and manager can exist as different rows in the same table with values in columns such as EmployeeID and ManagerID being used to help link the two rows. The PATH and PATHITEM functions are available in DAX to help provide a flattened view of these types of data structures inside calculated columns.

This was described to me once as a Pigs-Ear relationship. Not because it was considered a messy data modelling technique, but rather because of the way you might draw the line/arrow in any table relationship diagram. The arrow would look like an ear when drawn around the top corner of the box representing the table.

Other self-join relationships might be useful for finding rows from a table prior to or following a specific row to be used in comparison scenarios.

How to Join Tables Without a Relationship

Now that you understand some of the limitations of the built-in relationship mechanism, let's explore some of the functions provided in DAX to help us in scenarios where the standard relationship can't help.

The CROSSJOIN and GENERATE Functions

Probably my favorite and most used functions for joining tables are the GENERATE and CROSSJOIN functions. They have the feel and exhibit a behavior like an INNER JOIN in T-SQL when you combine them with the DAX FILTER function.

CROSSJOIN

The base syntax of the CROSSJOIN function is as follows.

```
CROSSJOIN ( <table>, <table> [, <table>]...)
```

The minimum requirement is for two tables to be passed, but you can add additional tables. The result is a cartesian effect in which every row from the first table is matched with every row from the second table. Then every row from joining the first two tables is matched with every row in the third table. This can quickly generate output that is a table with many rows.

The output of the CROSSJOIN function is a table. You can use this table as a calculated table if you are using a version of DAX that supports the use of calculated tables or as a table expression within a DAX calculation.

If the table you use in the first argument has 10 rows, and the second table also has 10 rows, the unfiltered output is 100 rows. If an additional table also has 10 rows, then the unfiltered output is now 1,000 rows. This is equivalent to the following T-SQL statement:

```
SELECT
      *
FROM TableA
      CROSS JOIN TableB
```

Or for those who are old school reading this book

```
SELECT * FROM TableA, TableB
```

These T-SQL statements yield the same result. Neither has a WHERE clause to filter the number of rows for the final output down so it returns the number of rows in TableA × the number of rows in TableB.

GENERATE

The base syntax for GENERATE is as follows:

```
GENERATE ( <table1>, <table3> )
```

This differs from CROSSJOIN in that only two tables can be used, otherwise the output is the full cartesian product of the tables passed as parameters. The two parameters must be different tables.

> **Note**   Use GENERATE, and not CROSSJOIN, when you are planning to use the
> FILTER function.

Let me use the following *unrelated* tables (Table 5-1 and 5-2) to demonstrate the behavior of the GENERATE function.

Table 5-1. *Dataset to Be Used as TableA*

ID	Make	Model	Value
1	Toyota	Corolla	10
2	Hyundai	Elantra	20
3	Ford	Focus	30

Table 5-2. *Dataset to Be Used as TableB*

Make	Model	Year	Index
Toyota	Corolla	2018	100
Toyota	Corolla	2018	200
Hyundai	Elantra	2019	300
Hyundai	Elantra	2019	400
Ford	Focus	2018	500
Ford	Focus	2019	600

Let's start with a simple query. A calculated measure to show the total number of rows from the result of the GENERATE function would look like this:

```
My Count of Rows =
    COUNTROWS(
        GENERATE('TableA','TableB')
        )
```

This returns the value of 18, which is the result of three rows from TableA multiplied by the six rows of TableB.

This calculation can be updated to apply a join criteria on the [Make] columns as shown in Listing 5-1.

Listing 5-1. Using GENERATE in a Calculated Measure with a Filter

```
My Count of Rows =
    COUNTROWS(
              GENERATE(
              'TableA',
              FILTER(
                       'TableB',
                       'TableA'[Make] = 'TableB'[Make]
                       )
              )
          )
```

The GENERATE function now uses a FILTER function as the second parameter that is applying the join criteria. This is the DAX equivalent of providing predicates in the ON or WHERE clauses in a T-SQL statement. The filter expression specifies that only rows between TableA and TableB that have matching values in the [Make] column should be returned. The result of this calculated measure is 6.

This could have been achieved using a standard DAX relationship. Let's now include a second column in the match requirement using [Model] as well as an additional criterion—that the value in the year column should be earlier than 2019.

The new calculation with the additional filter criterion is shown in Listing 5-2.

Listing 5-2. Using Additional Filter Criterion in a GENERATE Function

```
My Count of Rows =
    COUNTROWS(
              GENERATE(
              'TableA',
              FILTER(
                       'TableB',
                       'TableA'[Make] = 'TableB'[Make] &&
                       'TableA'[Model] = 'TableB'[Model] &&
                       'TableB'[Year] < 2019
                       )
              )
          )
```

The FILTER function allows for rules that are more sophisticated than those you can use with standard DAX relationships. The result of this calculated measure over the sample data should be 3. I could have used `'TableB'[Year] = 2018` in this dataset to arrive at the same result, but I wanted to highlight the ability to use an operator other than = for the matching criteria.

The same output could also be generated using a relationship and the following calculated measure:

```
My Count of Rows =
    CALCULATE(
        COUNTROWS('TableA'),
        'TableB'[Year] < 2019
        )
```

Unique Column Names

So far, the examples using GENERATE have all worked because they passed the output to the COUNTROWS function, which simply returned a count of the number of rows.

If you intend to use the output of a GENERATE function as a calculated table, you soon encounter a problem regarding column names. In DAX, column names in physical and virtual tables must be unique. The output of the GENERATE function includes all columns from all tables passed to the function. Inevitably some of these tables have the same column names.

If you try to create the following calculated table

```
My Table = GENERATE('TableA','TableB')
```

you encounter an error such as "The Column with the name [Make] already exists in the 'Table' table." In this example, the error refers to the clash over the [Make] column but also has an issue with the [Model] column that exists in both tables.

In T-SQL, you control the name and number of columns returned using the SELECT clause in a query. In DAX, you can use the SELECTCOLUMNS function to provide similar functionality. The primary use of this function is to allow you to partially select some columns from a table, but a side feature that's useful here is that it also allows you to rename, or alias, columns along the way.

Using the SELECTCOLUMNS function, you can rename the [Make] and [Model] columns in either table, or because you are using the = operator, you can choose to simply drop these columns from one of the tables, since their values should match and you only need to see one.

Listings 5-3 and 5-4 show two examples that use the SELECTCOLUMNS function to return an error-free result.

Listing 5-3. Using SELECTCOLUMNS with GENERATE

```
My Table =
    GENERATE(
            'TableA',
            SELECTCOLUMNS(
                'TableB',
                "My Make", [Make],
                "My Model", [Model],
                "Year", [Year],
                "Index", [Index]
            )
        )
```

Listing 5-4. Alternate Use of SELECTCOLUMNS with GENERATE

```
My Table =
    GENERATE(
            'TableA',
            SELECTCOLUMNS(
                'TableB',
                "Year", [Year],
                "Index", [Index]
            )
        )
```

In both examples, the second parameter of the GENERATE function is no longer just 'TableB'; rather, it has been wrapped using the SELECTCOLUMNS function. This function outputs a table that it shapes based on the rule you pass to it. In the first case, you instruct SELECTCOLUMNS to use 'TableB' and return four columns that are defined as <name>/<expression> pairs. This is how you can rename columns to avoid an error in the output of the GENERATE function.

The second example (Listing 5-4) simply omits the columns altogether for a cleaner effect and returns a table showing 18 rows, which is the full cartesian product of the two tables.

A more realistic requirement might be to combine this with the earlier example at Listing 5-4 to return a calculated table with unique column names for rows that matched on [Make] and [Model] for years before 2019.

The DAX for this is shown in Listing 5-5.

Listing 5-5. Combining GENERATE with FILTER and SELECTCOLUMNS

```
My Table =
    FILTER(
        GENERATE(
            'TableA',
            SELECTCOLUMNS(
                'TableB',
                "My Make", [Make],
                "My Model", [Model],
                "Year", [Year],
                "Index", [Index]
            )
        ),
        [Make] = [My Make] &&
        [Model] = [My Model] &&
        [Year] < 2019
    )
```

Here GENERATE is wrapped with the FILTER function to allow the calculation to define the rules used when you're matching rows between the tables. The inner SELECTCOLUMNS must return an alias for both the [Make] and [Model] columns in 'TableB' so the outer FILTER function can work with the table.

This highlights that the filtering doesn't take place during the join process; rather, it takes place over the single table output of the GENERATE function.

The T-SQL equivalent of this DAX statement including the renaming of the columns from 'TableB' is shown in Listing 5-6.

Listing 5-6. T-SQL Equivalent of DAX Example

```
SELECT
    A.[ID],
    A.[Make],
```

```
        A.[Model],
        A.[Value],
        B.[Make] AS [My Make],
        B.[Model] AS [My Model],
        B.[Year],
        B.[Index]
FROM TableA AS A
        INNER JOIN TableB AS B
            ON   A.Make = B.Make
            AND A.Model = B.Model
            AND B.Year < 2019
```

Both the DAX and T-SQL statements produce the output shown in Table 5-3.

Table 5-3. *The Output of the DAX and T-SQL Example*

ID	Make	Model	Value	My Make	My Model	Year	Index
1	Toyota	Corolla	10	Toyota	Corolla	2018	100
2	Hyundai	Elantra	20	Hyundai	Elantra	2018	300
3	Ford	Focus	30	Ford	Focus	2018	500

To remove the unnecessary [My Make] and [My Model] columns that were needed for the FILTER function, you can modify the calculation by wrapping the entire statement with another SELECTCOLUMNS statement. This is shown in Listing 5-7 using variables that can enhance readability.

Listing 5-7. Enhanced Version of Example Removing Unwanted Columns

```
My Table =

VAR ModifiedTableB =
    SELECTCOLUMNS(
        'TableB',
        "My Make", [Make],
        "My Model", [Model],
        "Year", [Year],
```

```
        "Index", [Index]
        )

VAR FilteredGENERATE =
    FILTER(
        GENERATE( 'TableA', ModifiedTableB ),
            [Make] = [My Make] &&
            [Model] = [My Model] &&
            [Year] < 2019
            )

RETURN
    SELECTCOLUMNS(
                FilteredGENERATE,
                "ID",[ID],
                "Make",[Make],
                "Model",[Model],
                "Year",[Year],
                "Index",[Index]
                )
```

This calculation introduces a SELECTCOLUMNS function at the final RETURN step that effectively drops the [My Make] and [My Model] columns by omitting these from the <name>/<expression> pairs.

If you need to, you can use more sophisticated expressions with the SELECTCOLUMNS function other than simply the name of a column.

Using GENERATE to Multiply Rows

Another way to take advantage of the GENERATE/FILTER functions is to use them to expand the number of rows you have in an existing table to solve a visualization problem.

A topic that arises from time to time on internet forums is how to take a table of entities that has a mixture of start/end dates and plot how many are active at any one time between their start/end dates.

Consider the dataset in Table 5-4 of hotel room occupancies. In this table, the date format is YYYY-MM-DD.

Table 5-4. *Dataset to Be Used as an Occupancies Table*

ID	Room	Check In	Check Out
1	101	2019-01-01	2019-01-02
2	102	2019-01-03	2019-01-04
3	103	2019-01-02	2019-01-05
4	101	2019-01-02	2019-01-05

To have a visual that shows how many of the rooms are occupied each night, one solution is to expand the data from the original table to carry a row that represents every room/night combination. For the occupancy with an ID of 4, the room is occupied for three nights (from January 2 to January 4), so the task of plotting this on a visual is easier if this table has three rows for ID = 4, rather than the single row in this table.

If there is a Date or Calendar table in the model with one row per day, including the days involved in the sample, you can use this to help create the rows you need. Listing 5-8 uses a dynamically created table as an alternative.

Listing 5-8. Using GENERATE with the CALENDAR Function

```
Table for Visual =
    SELECTCOLUMNS(
        FILTER(
            GENERATE(
                'Occupancies',
                CALENDAR("2019-01-01","2019-01-05")
                ),
            [Date] >= [Check In] &&
            [Date] < [Check out]
        ),
        "ID", [ID],
        "Room", [Room],
        "Date", [Date]
    )
```

This returns the three-column dataset in Table 5-5.

Table 5-5. *The Output of the Code from Listing 5-8.*

ID	Room	Date
1	101	2019-01-01
3	103	2019-01-02
4	**101**	**2019-01-02**
2	102	2019-01-03
3	103	2019-01-03
4	**101**	**2019-01-03**
3	103	2019-01-04
4	**101**	**2019-01-04**

As you can see, the row from the original Occupancy table with the ID of 4 now occurs three times in the new calculated table. This is the behavior you want and when plotted using the Ribbon Chart, it allows the visual to plot value for dates other than the original [Check In] and [Check Out] dates (Figure 5-1).

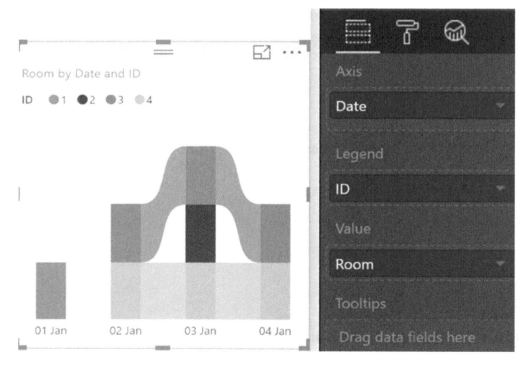

Figure 5-1. *Output from Listing 5-8 plotted using the Ribbon Chart visual*

You can apply this technique to a table that represents employees and is used with hire/leaving dates or any dataset that contains multiple rows of overlapping start/end values.

Before we dive in and break down the individual components, let's look at what the equivalent T-SQL statement might look like (Listing 5-9).

Listing 5-9. The T-SQL Equivalent of the Code in Listing 5-8

```
SELECT
      Occupancy.[ID],
      Occupancy.[Room],
      Dates.[Date]
FROM Occupancy AS O
      INNER JOIN Dates AS D
              ON  D.[Date] >= O.[Check In]
              AND D.[Date] < O.[Check Out]
```

Jumping back to the DAX calculation, let's break this apart to look at what each section is doing. Let's start with the parameters passed to the GENERATE function. The first parameter is straightforward in that you are simply passing your 'Occupancy' table. The second parameter is the DAX function CALENDAR. This handy function produces a single column table of dates between the two date parameters passed to it.

```
GENERATE(
    'Occupancies',
    CALENDAR("2019-01-01","2019-01-05")
),
```

In this case the two parameters are "2019-01-01" and "2019-01-05", which represent the values between January 1, 2019, and January 5, 2019. This results in a table with five rows and just a single column called [Date]. You could rename this column using SELECTCOLUMNS, but in this case, there is no column-name collision in the GENERATE function, so you can leave it as it is.

You could use a bigger range of dates, but doing so will not affect the result of the calculation, other than by possibly causing it to take slightly longer to generate data that will inevitably be ignored later in the calculation.

The output of the GENERATE function at this point is a 20-row table. This is the cartesian product of 4 rows from the 'Occupancies' table matched with all 5 rows from the table expression output of the CALENDAR function.

This is then wrapped with the FILTER function (Listing 5-10), which specifies rules involving multiple columns to help decide which rows to keep and which to disregard. You don't need dates that fall outside each occupancy for this requirement, but it could be interesting if you want to show when rooms are empty.

Listing 5-10. The FILTER Section from Listing 5-8

```
FILTER(
            GENERATE(
                'Occupancies',
                CALENDAR("2019-01-01","2019-01-05")
                ),
            [Date] >= [Check In] &&
            [Date] < [Check out]
        ),
```

The filter function accepts the table expression output of the GENERATE function as its first parameter. Filter rules are then applied to help reduce the dataset to only the rows you need.

This example highlights one of the other benefits of using the GENERATE function over standard table relationships. The FILTER function uses an operator other than = instead of using the >= and < operators to allow a match over a range of rows for each occupancy.

Finally, the FILTER function is wrapped inside a SELECTCOLUMNS function to control and name the columns returned to the calculated table.

Multiplying Rows Using a Numbers Table

A similar use of the GENERATE table to help multiply rows is to join to a table containing numbers. A *numbers table* is a general-purpose utility table that contains a single column of sequential numbers starting at zero and going up to a value that makes sense for your data model.

Let's say you want to expand the Table 5-6 to produce a copy of each row, with the number of copies controlled by the value in the Factor column.

Table 5-6. *Dataset to Be Used as Table1 Table*

Item	Factor
A	1
B	2
C	3

Listing 5-11 shows the DAX calculation table to expand this data.

Listing 5-11. Using GENERATESERIES to Create Rows

```
New Table =
    ADDCOLUMNS(
        FILTER(
            GENERATE(
                'Table1',
                GENERATESERIES(1,3)
                )
            ,[Factor] <= [Value]
        ),
        "New Col", [Factor] * [Value]
    )
```

This produces the output in Table 5-7.

Table 5-7. *Output from the Code in Listing 5-11*

Item	Factor	Value	New Col
A	1	1	1
B	2	1	2
B	2	2	4
C	3	1	3
C	3	2	6
C	3	3	9

To break apart this calculation, start with the inner GENERATE function. This is similar to the previous occupancy example, only instead of passing the CALENDAR function, it uses the GENERATESERIES function. This function generates a single row table with a range of whole numbers. This is the syntax for GENERATESERIES:

```
GENERATESERIES (<startValue>, <endValue> [, <incrementValue>])
```

The first and second parameters control the start and end values of the number range. This example creates a table with just three rows. The third parameter is optional and if it is not supplied, as in this example, it defaults to 1. The GENERATESERIES function is essentially creating a dynamic numbers table.

GENERATESERIES was introduced in October 2017 and is not currently available in all DAX environments. To replicate this functionally in earlier versions of DAX, use the CALENDAR function as follows:

```
GENERATE(
    'Table1',
    SELECTCOLUMNS(CALENDAR(1,3),"Value",INT([Date]))
)
```

The table expression output from the GENERATE function is passed to the FILTER function, which then applies the rule [Factor] <= [Value] to determine which rows to return.

Finally, the ADDCOLUMNS function is used as an alternative to SELECTCOLUMNS. In this example, you don't need to handle any column name collisions for the GENERATE function, so you can use ADDCOLUMNS to append a new column to all the existing output. This example also uses an expression that involves a calculation over multiple values to demonstrate how to include a more meaningful calculation if you need to.

Using GENERATE to Self-Join

The last example uses GENERATE to perform a self-join and uses a simplified 'Sales' table that shows dates on which a customer has made a purchase. For each purchase, the requirement here is to understand how long it has it been since that customer previously made a purchase.

The example uses the dataset in Table 5-8.

Table 5-8. *Dataset to Be Used as a Sales Table*

Customer	Purchase Date
1	2019-01-01
1	2019-01-07
1	2019-01-21
1	2019-01-25
2	2019-01-05
2	2019-01-12
2	2019-01-17
2	2019-01-22

The calculated table is shown in Listing 5-12.

Listing 5-12. The Calculated Table to Find Last Purchases Using Self-Join

```
New Table =

VAR SelfJoin =
    FILTER(
        GENERATE(
                'Sales',
                SELECTCOLUMNS(
                    'Sales',
                    "xCustomer",[Customer],
                    "xPurchase Date",[Purchase Date]
                )
        ),
        [xPurchase Date] < [Purchase Date] &&
        [Customer] = [xCustomer]
        )
```

```
VAR LastPurchases =
    GROUPBY(
            SelfJoin,
            Sales[Customer],
            Sales[Purchase Date],
            "Last Purchase",MAXX(
                                CURRENTGROUP(),
                                [xPurchase Date]
                                )
            )
RETURN
    ADDCOLUMNS(
        NATURALLEFTOUTERJOIN('Sales',  LastPurchases),
        "Days since last purchase",
                                    VAR DaysSince = INT([Purchase Date] -
                                    [Last Purchase])
                                    RETURN IF (
                                            NOT ISBLANK([Last Purchase]),
                                            DaysSince
                                            )
            )
```

The output of this calculation is Table 5-9, which shows the original table (Table 5-8) with two new columns. The new columns show the date the Customer made their previous purchase and a value that represents the number of days it has been since that purchase.

Table 5-9. *Dataset Showing Output of the Code in Listing 5-12*

Customer	Purchase Date	Last Purchase	Days Since Purchase
1	2019-01-01		
1	2019-01-07	2019-01-01	6
1	2019-01-21	2019-01-07	14
1	2019-01-25	2019-01-21	4
2	2019-01-05		
2	2019-01-12	2019-01-05	7
2	2019-01-17	2019-01-12	5
2	2019-01-22	2019-01-17	5

You can use the [Days Since Purchase] column in a variety of calculated measures to help you understand customer behavior such as the average time between purchases and the minimum and maximum gaps in time between purchases.

To understand the calculation, you can easily break the code in Listing 5-12 into three blocks made up of the two variables declarations and the final return.

The objective of the first variable is to create a table expression that joins every row from the eight-row 'Sales' table joined back to itself matching every row, but applying a filter that will match on CustomerID and for any previous purchase.

The T-SQL equivalent of this VAR statement is shown in Listing 5-13.

Listing 5-13. T-SQL Equivalent of the First Variable from Listing 5-12

```
SELECT
     L.Customer,
     L.[Purchase Date],
     R.Customer AS [xCustomer],
     R.[Purchase Date] AS [xPurchase Date]
INTO #SelfJoin
FROM Sales AS L
     INNER JOIN Sales AS R
          ON  R.Customer = L.Customer
          AND R.[Purchase Date] < L.[Purchase Date]
```

This produces a 12-row result that, for some transactions, finds multiple matches because at this point, the query is looking for ALL previous purchases. The very first purchase for each customer is missing because there are no prior purchases. The pattern of this query is similar to Listing 5-7 because it has used multiple columns in the join criteria, it uses an operator other than =, and finally, it uses SELECTCOLUMNS to rename columns to avoid a column-name collision.

At this point, the final RETURN statement could simply return the SelfJoin variable, which allows inspection of the data for correctness. The table expression produced by this variable is not intended as the final output; it is intended as a data-preparation step for the next variable.

The VAR LastPurchases statement summarizes the SelfJoin variable to a single line per [Customer] and [Purchase Date] combination. An aggregation column called "Last Purchase" is added, which finds the latest value in the [xPurchase Date] column. This identifies the date needed for the previous purchase.

Because the intention is to summarize a table expression rather than a physical table, the SUMMARIZE or SUMMARIZECOLUMNS functions are not available. However, the GROUPBY is available and can use the SelfJoin variable as its first parameter. The second and third parameters name the columns to group by. The final parameter is a DAX expression using the MAXX iterator function to find the latest value for each combination. The CURRENTGROUP() function is used to help the MAXX function make use of the table expression stored in the SelfJoin variable.

At this point the result is a six-row table (Table 5-10). This is not intended as the final result, nor does it carry any information about the first-ever purchase by each customer. This result set needs to be added back to the original data, which happens in the next step.

Table 5-10. *Dataset Output of First Variable from Listing 5-12*

Customer	Purchase Date	Last Purchase
1	2019-01-07	2019-01-01
1	2019-01-21	2019-01-07
1	2019-01-25	2019-01-21
2	2019-01-12	2019-01-05
2	2019-01-17	2019-01-12
2	2019-01-22	2019-01-17

The T-SQL equivalent of this statement is shown in Listing 5-14.

Listing 5-14. T-SQL Equivalent of Second Variable from Listing 5-12

```
SELECT
    [Customer],
    [Purchase Date],
    MAX([xPurchase Date]) AS [Last Purchase]
INTO #LastPurchases
FROM #SelfJoin
GROUP BY
    [Customer],
    [Purchase Date]
```

The last step is to take the table expression stored in the LastPurchases variable and join it back to the original table. This needs to be an outer semi join so you keep the row that represents the initial purchase by each customer.

The intention is to match every row from the original 'Sales' table with rows from the LastPurchases table expression, joining them where there is a match in both the [Customer] and [Purchase Date] columns. Rows in the 'Sales' table that cannot be matched to rows in the LastPurchases table expression need to be retained.

The GENERATE function does not easily provide "left join" functionality between tables. You can do this in a roundabout way by using multiple steps involving FILTER and further GROUPBY functions, or you can use the NATURALLEFTOUTERJOIN function to obtain the desired result.

In the final RETURN statement, the NATURALLEFTOUTERJOIN is passed two tables. The first is the original eight-row 'Sales' table. The second is the six-row table expression stored in the LastPurchases variables. The function checks every column in both tables and automatically matches any column that shares the same name and datatype, as well as the same lineage (more about this later).

```
NATURALLEFTOUTERJOIN('Sales', LastPurchases)
```

In this example, both the 'Sales' and LastPurchases tables passed to NATURALLEFTOUTERJOIN have [Customer] and [Purchase Date] columns that also share the same datatype. This returns a three-column table (Table 5-11) that still has one more requirement to fill.

Table 5-11. *Output of NATURALLEFTOUTERJOIN from Listing 5-12*

Customer	Purchase Date	Last Purchase
1	2019-01-01	
1	2019-01-07	2019-01-01
1	2019-01-21	2019-01-07
1	2019-01-25	2019-01-21
2	2019-01-05	
2	2019-01-12	2019-01-05
2	2019-01-17	2019-01-12
2	2019-01-22	2019-01-17

The final requirement is to have a column that carries a number that shows, for each customer purchase, how many days have passed since their last purchase. This is achieved by wrapping the NATURALLEFTOUTERJOIN function with an ADDCOLUMNS function (Listing 5-15).

Listing 5-15. The Final RETURN Statement from Listing 5-12

```
RETURN
    ADDCOLUMNS(
        NATURALLEFTOUTERJOIN('Sales', LastPurchases),
        "Days since last purchase",
                                    VAR DaysSince = INT([Purchase Date] -
                                    [Last Purchase])
                                    RETURN IF (
                                            NOT ISBLANK([Last Purchase]),
                                            DaysSince
                                    )
            )
```

The ADDCOLUMNS function appends a single column to the output of the NATURALLEFTOUTERJOIN function. The "Days since last purchase" parameter provides a name for the new column, while the last parameter starting with a nested VAR statement is a DAX expression that returns the difference between the [Purchase Date] and [Last Purchase] columns.

The INT function is used to convert the DateTime output of the subtraction to a whole number. You can use the DATEDIFF function at this point as long as the [Last Purchase] values are always the same or earlier than values in the [Purchase column].

This calculation returns the expected values for each row except for the customers' initial purchases. For these rows, it returns values such as 43,466 and 43,470, which could skew subsequent usage of this column in further calculations.

To address these high numbers, an IF function tests to see if the [Last Purchase] column carries a value. In the event that there is no value, it returns a blank, otherwise it returns the value showing the difference between the purchase dates.

For reference, the T-SQL equivalent of the final RETURN statement is shown in Listing 5-16.

Listing 5-16. T-SQL Equivalent of Final RETURN Statement from Listing 5-12

```
SELECT
      L.Customer,
      L.[Purchase Date],
      R.[Last Purchase],
      DATEDIFF(DAY,R.[Last Purchase],L.[Purchase Date]) AS [Days since last
      purchase]
FROM #Sales AS L
      LEFT OUTER JOIN #LastPurchases AS R
            ON  L.Customer = R.Customer
            AND L.[Purchase Date] = R.[Purchase Date]
```

A simpler alternative version of the calculated table that produces the same result but doesn't use a self-join technique is shown in Listing 5-17.

Listing 5-17. Simpler Alternative for Listing 5-12

```
New Table =
    ADDCOLUMNS(
        ADDCOLUMNS(
            'Sales',
            "Last Purchase", MAXX(
                            FILTER(
                                'Sales',
                                [Customer] = EARLIER([Customer]) &&
```

117

```
                                        [Purchase Date] < EARLIER([Purchase
                                        Date])
                                        ),
                                    [Purchase Date]
                                    )
                        ),
    "Days since last purchase", IF(
                                NOT ISBLANK([Last Purchase]),
                                DATEDIFF([Last Purchase], [Purchase
                                Date],DAY)
                                )
    )
```

NATURALINNERJOIN and NATURALLEFTOUTERJOIN

The NATURALINNERJOIN function allows you to join two tables using the following syntax:

```
NATURALINNERJOIN ( <leftTable>, <rightTable>)
```

This function matches every row from the first table with every row from the second table that has matching values in any column that shares the same column name and datatype.

If a row from either table cannot be matched to a row in the other table, it is dropped from the result. The additional requirement is that both tables must be derived from the same physical source table. This is known as having the *same lineage*.

Once matching rows are found, the columns used in the match are returned once along with any additional columns from the left and right tables that were not used as part of the matching exercise.

The NATURALLEFTOUTERJOIN function has the same two-parameter requirement, only it keeps rows from the table passed as the first parameter that find no matching rows from the table passed as the second parameter.

As the names suggest, these functions are like the INNER JOIN and LEFT OUTER JOIN statements in T-SQL.

Lineage

A lineage requirement exists for both the NATURALINNERJOIN and NATURALLEFTOUTERJOIN functions, which tends to limit the number of scenarios in which they might be useful.

In the WideWorldImportersDW dataset, the 'Fact Sale' table contains a column called [City Key], which uses the whole number datatype. The 'Dimension City' table has a column with the same name and datatype, but because these are separate tables, the following DAX produces an error.

```
New Table = NATURALINNERJOIN('Fact Sale','Dimension City')
```

Wrapping the NATURALINNERJOIN with a FILTER in the same way you did earlier with GENERATE does not get around this error. It is the same with NATURALLEFTOUTERJOIN. The most common use of these functions is similar to the previous example to look for previous purchases as shown at Listing 5-17.

Typically a single table is used as a starting point in a multistatement DAX calculation that involves variables. One or more table expressions are generated from the source table and stored in variables. These table expressions may be summarized or filtered to manipulate the source data to then join back to the original table. In this case, the DAX engine knows that although you may be working with a column that has been aggregated or renamed many times, it can still trace the value back to the same source table involved in the join.

One way to work around this limitation is to use the SELECTCOLUMNS function. Tables 5-12 and 5-13 demonstrate this.

Table 5-12. *Dataset to Be Used as the Table1 Table*

ID	Value1
A	1
B	2
C	3
D	4

Table 5-13. *Dataset to Be Used as the Table2 Table*

ID	Value2
C	3
D	4
E	5
F	6

The following calculation to join these tables fails with an error.

```
New Table = NATURALINNERJOIN('Table1','Table2')
```

This is despite the calculation meeting two of the requirements of having a column in each table using the same name and datatype. NATURALLEFTOUTERJOIN produces the same error. To get around this, use SELECTCOLUMNS as shown in Listing 5-18. Table 5-14 shows the output from this code.

Listing 5-18. Calculated Table Using NATURALINNERJOIN

```
New Table =
VAR LeftTable  =
     SELECTCOLUMNS(
                     'Table1',
                     "ID",[ID] & "",
                     "Value1",[Value1]
                     )
VAR RightTable =
     SELECTCOLUMNS(
                     'Table2',
                     "ID",[ID] & "",
                     "Value2",[Value2]
                     )
RETURN
     NATURALINNERJOIN(
                 LeftTable,
                 RightTable
                 )
```

Table 5-14. *The Output of the Code in Listing 5-18*

ID	Value1	Value2
C	3	3
D	4	4

An approach to take here is to append an empty text value to the column name of each column you intend to use to join by the NATURALINNERJOIN function. The same technique works for NATURALLEFTOUTERJOIN and matches on multiple columns that share the same name and datatype. There are probably good reasons why the lineage requirement exists for these functions, particularly in terms of performance over larger datasets, so be mindful of this when you are using this technique.

The output of the same query using NATURALLEFTOUTERJOIN is shown in Table 5-15.

Table 5-15. *The Output of Code from Listing 5-18 Using NATURALLEFTOUTERJOIN*

ID	Value1	Value2
A	1	
B	2	
C	3	3
D	4	4

UNION

The join functions covered so far allow tables to be joined horizontally. If you need to join two or more vertically, you can use the UNION function. This operates the same way UNION ALL works in T-SQL. The syntax for UNION is as follows:

```
UNION ( <table1>, <table2> [, <tableN>...])
```

The function expects each table passed to have the same number of columns, otherwise, an error is produced. SELECTCOLUMNS and ADDCOLUMNS are useful functions to help you shape tables to the point where they have the same number of columns for UNION to work.

An effective use case for UNION is in conjunction with NATURALLEFTOUTERJOIN over a base table that is missing data. You can use the NATURALLEFTOUTERJOIN to help identify the gaps and then generate a table with dummy values that can be joined vertically back to the base table using the UNION function for a more complete table.

If you use the UNION function with the Table1 and Table2 datasets (see Tables 5-12 and 5-13) from the previous example, the output is shown in Table 5-16.

Table 5-16. *The Output of the UNION of Table1 and Table2*

ID	Value1
A	1
B	2
C	3
D	4
C	3
D	4
E	5
F	6

Note that the rows from Table1 and Table2 that are identical still return as separate rows and you don't have an option to perform a DISTINCT style T-SQL during the UNION. If you want to do this, you can wrap the UNION with a DISTINCT function to remove duplicate rows:

```
Table =
    DISTINCT(
        UNION('Table1',Table2)
        )
```

Columns are matched in the order in which they appear in the tables passed to the UNION function. If there is a difference in the datatype for a column between columns that are being connected, the DAX engine falls back to the datatype that can carry both sets of data; for example, if a whole number column is aligned with a text column, the resulting column has a datatype of Text.

It's also possible to union a table onto itself as a technique to quickly double or triple your source table. The following code returns a table that has three times the number of rows as Table1.

```
Union Table = UNION(Table1,Table1,Table1)
```

LOOKUPVALUE

The last function I cover in this chapter is one that may be more familiar to power users of EXCEL than T-SQL. The function is not so much a tool to join tables as it is a function to allow you to retrieve values from other tables without using a DAX relationship.

The syntax for LOOKUPVALUE is as follows:

```
LOOKUPVALUE ( <resultColumnNAme>, <searchColumnName>, <searchValue>

[, <searchColumnName>, <searchValue>]...)
```

The first parameter specifies the value to be returned by the function. When you use this in a calculated column, this is the value that ultimately shows in the new column. The column can be any column from any table including the table you may be adding to the calculated column.

The second and third parameters are <search>/<value> pairs. The <searchColumnName> parameter needs to be a column from the same table as the first parameter. The <searchValue> carries either a column to be used for values to search for, or it can be a hardcoded value.

The function needs to return a single result, so the LOOKUPVALUE function generates an error if it finds more than one distinct value. A common use of LOOKUPVALUE is between a table that might normally exist on the many side of a one-to-many relationship and the LOOKUPTABLE being used to search the table on the one side of the relationship for a value.

An example that uses the WideWorldImportersDW data might be to add a column to the 'Fact Sales' table that carries a value that it finds from the 'Dimension City' table. This does not need a relationship to be defined between these tables.

Listing 5-19 adds a column that shows the [Continent] for each sale.

Listing 5-19. Calculated Column Using LOOKUPVALUE Function

```
New Column =
     LOOKUPVALUE(
                      -- Value to be returned for this new column --
                      'Dimension City'[Continent],
                      -- Column to search --
                      'Dimension City'[City Key],
                      -- Value in column to search --
                      'Fact Sale'[City Key]
                      )
```

The first parameter 'Dimension City'[Continent] specifies the column that is used to find a value that eventually returns to the new calculated column.

The row used to return the value for [Continent] is determined by the next two parameters. 'Dimension City'[City key] tells LOOKUPVALUE this is the column you want to use when you are looking for rows, and the 'Fact Sale'[City Key] tells LOOKUPVALUE the exact value to look for. This case uses the value from the [City Key] column in the 'Fact Sale' table.

The <searchValue> parameter can be hardcoded or the result of a calculation. If it is hardcoded, it means you have the same value in every row of the new column.

Additional <searchColumnName> and <searchValue> sets of criteria can be passed to the LOOKUPVALUE function, which might provide more flexibility over a standard relationship, however this is likely to be inefficient over larger datasets on top of the requirement to ensure the LOOKUPVALUE function returns a single value for each search.

CHAPTER 6

Filtering

Implicit Filtering

Filtering data is inevitable in modern reporting solutions. It is rare these days when you need to create a report that simply counts the rows or performs simple calculations over every row in a table.

One of the many excellent features of interactive tools such as Power BI is how they can apply filters interactively over the underlying data. If you click around on visuals in a report page, other visuals react and respond dynamically to show you new values representative of the selection you just made.

This is mostly implicit filtering that happens automatically due to rules in the underlying data model. Selecting from a slicer, filter, or visual not only filters the table the data has comes from but propagates down through to tables that have relationships defined to the table you have just filtered, and not just to tables directly related, but to intergenerational tables, as long as they are on the many side of one-to-many relationships.

Filters *can* propagate from the many side through to the table on the one side as long as the cross-filtering property of the relationship is set to both.

For simple examples, it's possible to meet your requirements by importing data into a data model and creating appropriate relationships. Implicit measures can be used in your report, meaning you can get away with writing no code and still produce a useful report.

Using the WideWorldImportersDW example to help you understand, let's focus on three tables and walk through how filtering works using standard relationships and then look at how you can use some of the filter functions provided in DAX for extra control and flexibility.

© Philip Seamark 2018
P. Seamark, *Beginning DAX with Power BI*, https://doi.org/10.1007/978-1-4842-3477-8_6

The three tables are 'Dimension Date', 'Dimension Stock Item', and 'Fact Sale', with the relationships in Figure 6-1 defined in the data model. 'Dimension Date' has a one-to-many relationship to 'Fact Sale', with the 'Dimension Date' table on the one side. The column on the 'Fact Sale' table is [Invoice Date Key], whereas the column used in the 'Dimension Date' table is [Date]. Both these columns use the Date datatype.

The 'Dimension Stock Item' table also has a one-to-many relationship to the 'Fact Sale' table with [Stock Item Key] used as the column for each table.

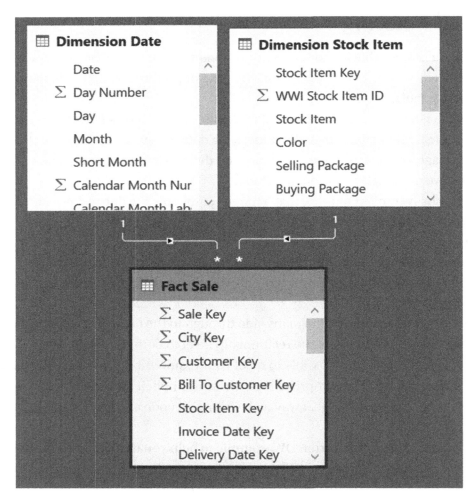

Figure 6-1. *The relationships between three tables in the model*

Tip   In the Relationship View, try to put the tables on the one side of one-to-many relationships above the tables on the many side. This helps to visually reinforce the trickle-down effect of filters. Selecting a column in a table on the one side filters rows in the table on the many side, but not the other way around.

This is a common pattern in data modelling for analytics and reporting. The table on the one side is generally a list of items you might like to slice and dice from the rows of the table on the many side. As the table name suggests, these are often referred to as *dimension tables*.

Dimension tables typically have one row per entity that represents the lowest level of detail to group by. The column that represents the lowest grain should be unique and contains the value that appears in your related table on the many side.

The tables on the many side can have multiple rows with the same value in the column used in the relationship. These are often referred to as fact tables. Hopefully your 'Fact Sales' table has many rows for any given date. Equally, it's possible and quite normal for the table on the many side to have no rows that match a column on the one side. If trading doesn't take place on weekends or holidays, there may not be rows for these days, and this is important to remember when you design your data model.

Fact tables often represent actions, events, or transactions, and usually they have a time component. A good rule of thumb to help decide which data should be shaped into a Fact table versus a Dimension table is this: if you would like to count, sum, or show a trend of the data over time, then consider using a fact table. If the data is how you might like to slice and dice, then it is probably a suitable candidate to use as a dimension table.

In the WideWorldImportersDW dataset, the lowest grain for the 'Dimension Date' table is a specific day and does not contain columns or rows that represent hours, minutes, or seconds.

Additional columns in the 'Dimension Date' table allow you to group days in useful ways, such as by calendar month. These additional columns do not have to be unique and can be customized to suit your organization's reporting requirements.

The 'Dimension Date' table in this dataset allows rows to be grouped into calendar and fiscal ranges such as months and years. Additional columns provide alternative formatting such as [Short Month] (Jan, Feb, Mar, . . .), and the [Month] column (January, February, March, . . .).

Any column used from 'Dimension Date' in any part of your report automatically sends filtering information to the related 'Fact Sale' table via the columns used in the relationship. There is no need to only use the columns defined in the relationship in slicers, filters, or in the axis areas of your visuals.

Starting with a very simple reporting requirement to show a count of the number of sales by invoice date, we can add the following calculated measure to the data model:

```
Count of Sales = COUNTROWS('Fact Sale')
```

When this measure is dropped onto the reporting canvas using a table visual, it shows a value of 228,265. This happens to be the same number of rows as are in the 'Fact Sale' table. No filtering has taken place in the calculation of this value; the COUNTROWS function has simply run a single pass over the 'Fact Sale' table counting every row.

If the [Invoice Date Key] field from the 'Fact Sale' table is added to the same visual—note that this is the field from the 'Fact Sale' table, not the 'Dimension Date' table—you should see what appears in Figure 6-2.

Invoice Date Key	Count of Sales
Tuesday, 1 January 2013	89
Wednesday, 2 January 2013	207
Thursday, 3 January 2013	215
Friday, 4 January 2013	146
Saturday, 5 January 2013	114
Monday, 7 January 2013	330
Tuesday, 8 January 2013	126
Wednesday, 9 January 2013	191
Thursday, 10 January 2013	258

Figure 6-2. *Sample of table visual using [Invoice Date Key] and implicit measure*

Figure 6-2 shows the first few rows of a result set that has 1,069 rows. Each cell in the [Count of Sales] column has effectively run the calculated measure, but now an implicit filter context is being applied (also known as *query context*).

This means the value of 89 in the top row of the Count of Sales column has effectively run the following calculation.

```
Count of Sales = COUNTROWS(
                         FILTER('Fact Sale',
                             'Fact Sale'[Invoice Date
                             Key]=DATE(2013,1,1)
                             )
                         )
```

The DAX engine has dynamically applied a filter condition to the original calculation, which is specific to that specific cell.

Note   Remember that every value shown in your report is the result of its own unique calculation and does not rely on the order or output of other calculations. This includes totals and subtotals.

The row for Wednesday, January 2, 2013, returned a value for a Count of Sales measure of 207. This is the result of DAX applying a slightly different automatic filter to the original calculation. In this case, the filter applied has the effect of asking the calculation engine to count every row in the 'Fact Sales' table that meets the criteria that the value in the [Invoice Date Key] table must equal January 2, 2013.

This first example used the [Invoice Date Key] from the 'Fact Sale' table. Now let's use the [Date] column from a related table instead and have a look at what happens (Figure 6-3).

Date	Count of Sales
Tuesday, 1 January 2013	89
Wednesday, 2 January 2013	207
Thursday, 3 January 2013	215
Friday, 4 January 2013	146
Saturday, 5 January 2013	114
Monday, 7 January 2013	330
Tuesday, 8 January 2013	126
Wednesday, 9 January 2013	191
Thursday, 10 January 2013	258
Friday, 11 January 2013	274

Figure 6-3. *Sample of table visual using [Date] column from 'Dimension Date' table*

The calculations produce the same result, only this time we used a column from a different table. The DAX engine automatically applies a filter to every calculation in the Count of Sales column (except, in this case, for the very bottom total). The difference here is that the filtering is taking place in the 'Dimension Date' table, which then makes its way down through the one-to-many relationship and is applied to the data in the 'Fact Sale' table.

To generate a value of 89 for the top row, the DAX engine knows you want to filter the data for rows associated with the January 1, 2013. The [Date] column is unique in the 'Dimension Date' table and is also the column specified in the active relationship between 'Dimension Date' and 'Fact Sale'.

Information contained in the relationship is required because you are using columns from two tables to generate the value for the calculated measure. So, the DAX engine finds every row in the 'Fact Sale'[Invoice Date Key] column that has an exact match with the filtered value from the 'Dimension Date'[Date] column. It's these rows that are used by the [Count of Sales] calculated measure to generate a value for the relevant cells.

A good question to ask at this point is why you should use the column from the 'Dimension Date' table when it produces the same result as using the [Invoice Date Key] column from the actual 'Fact Sales' table. Surely it is more efficient to use a column from the same table to filter the rows in which the data is being used in the calculations?

Some advantages of using filters from related tables include the following:

> *Dimension tables can be used to filter multiple fact-style tables.* This example contains a single fact table. If you also had report visuals that used data from other fact tables, such as 'Fact Purchase' or 'Fact Order', these probably have relationships to the same 'Dimension Date' table. Any slicer or filter based on a column from the 'Dimension Date' table automatically propagates down through multiple one-to-many relationships to be applied to calculations using columns in each of the related fact tables.

> *Dimension tables allow you to include columns that group rows together in meaningful ways.* The 'Date' table contains a unique row for every day. A [Calendar Year] column can carry the value of "2019" for all 365 rows in 'Dimension Date' that have a value between January 1, 2019, and December 31, 2019. Equally, columns in the 'Dimension Date' table can carry values that help identify and filter rows for specific months, quarters, working

days, holidays, specific events, and so on. Storing these "useful columns to group by" in a single table makes for a more efficient and easier-to-maintain data model.

Dimension tables typically guarantee one row per entity that they represent and are not prone to the same gaps and duplication that you may see in a fact table. In the simple case I have used here, there is no data for Sunday, January 6, 2013. This probably reflects the way this organizations works. By using the [Date] column from the 'Dimension Date' table, you can use the guaranteed availability of *all* days to your advantage in some calculations— particularly time intelligence DAX functions designed to assume the column being used contains no gaps.

In Figure 6-4, both visuals use the 'Dimension Date' [Date] column rather than the 'Fact Sale' [Invoice Date Key] column. These both clearly show no rows for January 6. This gap could be important and easily lost if you were using the [Invoice Date Key] column in the row header or axis.

Figure 6-4. *Table and bar chart visuals using the [Date] column with a gap for a day with no data*

Let's go back to the original example for just one more tweak. Rather than using the 'Dimension Date'[Date] column in the visual, let's use another column from the 'Dimension Date' table along with the [Count of Rows] calculated measure. Remember the [Date] column is guaranteed to be unique because it is being used as the column on the one side of a one-to-many relationship, so for every row in the example used so far, only one value is being passed through the relationship to the 'Fact' table as part of the filter context.

This time let's use the [Calendar Year] column from the 'Dimension Dates' table in the visual. This should produce the result shown in Figure 6-5.

Calendar Year	Count of Sales
2013	60,968
2014	65,941
2015	71,898
2016	29,458
Total	**228,265**

Figure 6-5. *The table visual using the [Calendar Year] field and implicit measure*

Figure 6-5 shows five values being generated by the [Count of Sales] calculated measure. Four of these are being filtered by the [Calendar Year] column, while the very bottom value of 228,265 is not affected by any filtering. The result of 60,968 for the top row starts by applying a filter over the 'Dimension Date' table looking for all rows in the which the [Calendar Year] column has a value of 2013. This should find exactly 365 rows that the DAX engine then uses to find every associated [Date] value. This group of [Date] values filters through the one-to-many relationship down to the 'Fact Sale' table to help identify which rows can be used by the calculated measure.

The DAX used in the [Count of Sales] calculated measure only mentions the 'Fact Sale' table, which is passed to the COUNTROWS function. There is no mention in the calculation about specific columns, how matching is to take place, or anything about other tables in the data model. All this filtering happens implicitly, with a little bit of help from a predefined relationship between two tables. DAX provides a FILTER function, but it is not needed to satisfy this requirement.

Explicit Filtering

All the filters in the preceding example Figure 6-5 were used in implicit filtering, meaning no DAX filter functions were used in the calculation itself. This allows basic reports to be created, but inevitably, you need more sophisticated filtering logic to meet more sophisticated reporting requirements.

DAX provides several filter functions and this section examines some common and useful examples using these functions. Filter functions in DAX not only allow you to provide additional layers of filtering to the implicit filtering for calculations, but they also allow you to specifically override the current filter context, allowing your calculations to use data from rows they might not otherwise have access to.

The FILTER Function

The first filter function we look at is the FILTER function. The syntax for this function is as follows:

```
FILTER ( <table>, <filter expression> )
```

The first parameter is a reference to the table you would like the filter to be applied to. The second parameter needs to be a valid DAX expression that results in a Boolean true or false that is logically applied to every row in the table used as the first parameter. The output of the FILTER function is a table. This is important to know when you're using the FILTER function nested inside other DAX functions. It may be used as parameter for any function that accepts a table as a reference.

Sticking with the [Count of Sales] measure, let's add the FILTER function with a simple expression to see what happens to the calculation.

Consider the calculated measure in Listing 6-1.

Listing 6-1. Using the FILTER Function with COUNTROWS

```
Count of Sales (10 or more) =
                    COUNTROWS(
                        FILTER(
                            'Fact Sale',
                            'Fact Sale'[Quantity] > 10
                            )
                        )
```

The difference between this and the earlier calculation is this version uses a FILTER function in place of the original 'Fact Sale' table reference. The result of this change is shown in Figure 6-6.

Calendar Year	Count of Sales	Count of Sales (10 or more)
2013	60,968	30,588
2014	65,941	33,113
2015	71,898	35,550
2016	29,458	15,436
Total	**228,265**	**114,687**

Figure 6-6. *Adding a [Count of Sales (10 or more)] measure to a table visual*

The new calculated measure is added to the table visual as a third column. To understand how we arrived at a value of 30,588 for the top row, let's look at what is happening. In this example, there are two layers of filtering taking place. The first layer is the implicit filter (query context), which is being driven from the [Calendar Year] column. The second layer is the explicit rule used in the FILTER function for Listing 6-1 over the 'Fact Sale'[Quantity] column.

The implicit filter context restricts the number of rows visible to the COUNTROWS function to 60,968 rows. Then for each of these rows, a Boolean test is applied to determine if the value from the [Quantity] column is higher than 10. This test effectively discards any of the remaining 60,968 rows that have a value lower than 11. The result is a DAX table returned from the FILTER function to the COUNTROWS function that happens to have 30,588 rows to be counted.

The key thing to remember is filtering is ordered, applied in layers, and is not a single operation. This allows for additional flexibility when you are creating DAX for more sophisticated reporting requirements.

You can extend the DAX used as the filter expression to include more specific requirements, as shown in Listing 6-2.

Listing 6-2. Calculated Measure Using Filter Criteria from a Related Table

```
Count of Sales (10 or more) =
                    COUNTROWS(
                    FILTER(
                        'Fact Sale',
                        'Fact Sale'[Quantity] >= 10
                        && RELATED('Dimension Stock Item'[Size]) = "M"
                        )
                    )
```

This example applies an additional rule that limits the number of rows from the 'Fact Sale' table still further to use the implicit filter (query context) being passed down from the 'Dimension Date' table due to the use of the [Calendar Year] column in the visual. Then in addition to the original explicit filter requirement to only consider rows that have a value of 10 or higher, an extra rule has been added that applies a filter to another related table. Because a relationship exists between 'Dimension Stock Item' and 'Fact Sales', this filter requirement is also applied to ultimately restrict the number of rows from 'Fact Sales' that are eventually passed to the COUNTROWS function to be counted.

The result is now a smaller value for the calculated measure in the third column of Figure 6-7.

Calendar Year	Count of Sales	Count of Sales (10 or more)
2013	60,968	1,534
2014	65,941	1,671
2015	71,898	1,793
2016	29,458	708
Total	**228,265**	**5,706**

Figure 6-7. *Output using the code from Listing 6-2*

Using the FILTER function in this way allows you to apply additional restrictions on top of any implicit filter (query context) that happens to be in effect. This means you can create calculations that have some rules hardcoded into the calculation but can still use the measures in many visuals, each with their own implicit filter context that can be combined with the explicit filters to generate useful dynamic results.

Overriding Filters

So far, the filtering has been cumulative, meaning different layers of filtering have simply been discarding rows at various points in the calculation. In DAX, there are filter functions that provide the ability to override filtering applied by the implicit filter layer. These are the functions we will look at that allow this:

- CALCULATE and CALCULATETABLE
- ALL
- ALLEXCEPT
- ALLSELECTED

There are several scenarios in which these functions can be useful. Examples include their ability to perform period comparisons, running totals, and percentage of total calculations, to name a few. In each of these scenarios, a calculation may need access to values stored in rows that have been disregarded by the implicit filter.

Percentage of Total

Starting with a percentage of total measure, you can use the count of rows by year example to demonstrate how to override the implicit filter context. By default, any calculated measure used in a visual has a layer of automatic filtering applied to it driven by other columns used in the same visual.

```
Count of Sales = COUNTROWS('Fact Sale')
```

The [Count of Sales] calculated measure, when used in the same visual as the 'Dimension Date'[Calendar Year] column, has a layer of implicit filtering that means the COUNTROWS function can only consider a portion of the actual rows from 'Fact Sales'.

If the requirement is to understand what proportion each year is of the overall total, you need a way to instruct the DAX engine to ignore the implicit filter context. DAX provides CALCULATE and CALCULATETABLE to help manage this.

Figure 6-8 demonstrates this using the following calculated measure:

```
Count of all Sales = CALCULATE(
                        COUNTROWS('Fact Sale'),
                    ALL('Fact Sale')
                    )
```

Calendar Year	Count of Sales	Count of all Sales
2013	60,968	228,265
2014	65,941	228,265
2015	71,898	228,265
2016	29,458	228,265
Total	**228,265**	**228,265**

Figure 6-8. *Output using the ALL function*

Let's break this down and explain the calculation in more detail.

First, the inner COUNTROWS function is the same as what was used in the [Count of Sales] calculated measure. It counts the number of all rows in the 'Fact Sale' table that it can see. This is normally a version of the table filtered by the query context. In this case, the COUNTROWS function is wrapped inside a CALCULATE function.

The CALCULATE function has the following syntax:

```
CALCUALTE ( <expression>, <filter1>, <filter2> ... )
```

The function executes the expression provided in the first parameter in the context of the filters supplied by the additional parameters. As the syntax for the CALCULATE function suggests, there can be any number of filter expressions. In the [Count of all Sales] example, the parameter being passed to the CALCULATE function is the ALL function. The effect of using the ALL function here is to remove any implicit filter context that may be in effect on the 'Fact Sale' table, which is the table specified as the parameter of the ALL function.

Note   The use of CALCULATE and ALL in this way removes all filter context including any hardcoded explicit filter rules you may have used elsewhere such as a Report, Page, or Visual filter.

This now allows for a calculation to be written that combines the [Count of Sales] with [Count of all Sales] to generate a value that shows a percentage of total for each row. The output of this calculation is shown in Figure 6-9.

```
Count of Sales % =
            DIVIDE(
                [Count of Sales],
                [Count of all Sales]
                )
```

Calendar Year	Count of Sales	Count of all Sales	Count of Sales %
2013	60,968	228,265	26.71%
2014	65,941	228,265	28.89%
2015	71,898	228,265	31.50%
2016	29,458	228,265	12.91%
Total	228,265	228,265	100.00%

Figure 6-9. Output with an additional calculated measure

In this example in Figure 6-9, the table passed to the ALL function was 'Fact Sales.' The function removed any implicit filter context from the 'Fact Sales' table, which was the query context being generated by the 'Dimension Date'[Calendar Year] column. If the 'Dimension Date' table was used in place of 'Fact Sale,' the filter context would instead be removed from the 'Dimension Date' table. This subtle change doesn't end up changing the values of the visual used in this example.

The net effect of the overall filtering still passes the same rows to the COUNTROWS function. There may be a small difference in performance, but we look at that in Chapter 8 when I cover optimizing DAX.

If you include some additional columns to the table visual, this will demonstrate the difference when you specify which table to use in the ALL function with calculated measures.

Consider the two calculated measures in Listings 6-3 and 6-4.

Listing 6-3. Calculated Measure Using ALL with the 'Dimension Date' Table

```
Count of all (Dim Date) Sales =
                CALCULATE(
                    COUNTROWS('Fact Sale'),
                    ALL('Dimension Date')
                    )
```

Listing 6-4. Calculated Measure Using ALL with the 'Fact Sale' Table

```
Count of all (Fact Sales) Sales =
                CALCULATE(
                    COUNTROWS('Fact Sale'),
                    ALL('Fact Sale')
                    )
```

The only difference between these calculated measures is the table used in the ALL function. These yield the same results via two separate paths when you use them in a visual with just the 'Dimension Date'[Calendar Year] column. However, if you introduce a filter from a table other than 'Dimension Date', you will see a difference. This case (shown in Figure 6-10) uses the 'Fact Sale'[Package] column, in which all rows have just one of the following four values: Bag, Each, Packet, or Pair.

Calendar Year	Package	Count of Sales	Count of all (Dim Date) Sales	Count of all (Fact Sales) Sales
2013	Bag		1,036	228,265
2014	Bag		1,036	228,265
2015	Bag		1,036	228,265
2016	Bag	1,036	1,036	228,265
2013	Each	58,439	217,808	228,265
2014	Each	63,181	217,808	228,265
2015	Each	68,943	217,808	228,265
2016	Each	27,245	217,808	228,265
2013	Packet	1,393	5,209	228,265
2014	Packet	1,561	5,209	228,265
2015	Packet	1,609	5,209	228,265
2016	Packet	646	5,209	228,265
2013	Pair	1,136	4,212	228,265
2014	Pair	1,199	4,212	228,265
2015	Pair	1,346	4,212	228,265
2016	Pair	531	4,212	228,265
Total		**228,265**	**228,265**	**228,265**

Figure 6-10. *Output using code from Listings 6-3 and 6-4*

For each cell/computation in the third, fourth, and fifth columns of Figure 6-10, the calculated measure starts with an automatic filter context that is relevant for the row in question. The [Count of Sales] measure doesn't remove any of the implicit filter context, so the values reflect the true number of rows in the 'Fact Sale' table that happen to match the [Calendar Year] and [Package] columns.

However, for the [Count of all (Dim Date) Sales] calculated measure, the only filter context being removed is any that exists on the 'Dimension Date' table and not any filter context being applied to the 'Fact Sale' table. This explains why you see different values between the rows where the value for [Package] is Bag and rows with a different value for [Package].

What is happening here is the implicit filter being driven by the [Calendar Year] column is being removed, while the implicit filter being driven by the [Package] column is not being removed.

For the [Count of all (Fact Sales) Sales] calculated measure, the ALL function is clearing all implicit filters from the 'Fact Sales' table, which includes any implicit filters being passed down from related tables; therefore, the COUNTROWS function can now use every row in the 'Fact Sale' table to generate a value.

The implied filter context is not just driven from other columns used in the same visual. Implied filter context can be driven from filters external to the current visual. These can take the form of selections made to slicers on the same report, or other non-DAX-based external filter settings.

Explicit Filters

Functions that allow the removal of implicit filter context have no effect on explicit filters. This means that if you use a specific filter rule in a calculated measure, this filter condition cannot be removed by a downstream calculation.

Consider the two calculated measures in Listings 6-5 and 6-6.

Listing 6-5. Calculated Measure Applying Filter to [Quantity] Column

```
Count of Sales (10 or more) =
                    COUNTROWS(
                        FILTER(
                            'Fact Sale',
                            'Fact Sale'[Quantity] >= 10
                        )
                    )
```

Listing 6-6. Calculated Measure Showing the effect of the ALL Function with the Calcuated Measure from Listing 6-5

```
Count of all (Explicit) Sales =
                    CALCULATE(
                        [Count of Sales (10 or more)],
                        ALL('Fact Sale')
                    )
```

When the calculation is used in the table visual, you get the output shown in Figure 6-11.

Calendar Year	Count of Sales	Count of all (Fact Sales) Sales	Count of Sales (10 or more)	Count of all (Explicit) Sales
2013	60,968	228,265	30,588	114,687
2014	65,941	228,265	33,113	114,687
2015	71,898	228,265	35,550	114,687
2016	29,458	228,265	15,436	114,687
Total	228,265	228,265	114,687	114,687

Figure 6-11. *Output using code from Listings 6-5 and 6-6*

In Figure 6-11, the fourth column uses the [Count of all (Explicit) Sales] calculated measure. The value shown in each cell is the result of a mixture of implicit filter context and explicit filtering. The value of 30,588 represents the number of rows from 'Fact Sales' that belong to the [Calendar Year] of 2013, which is the implicit filter context, along with the additional restriction to only count rows where the 'Fact Sale'[Quantity] is greater than or equal to 10. You end up with five different values, including the Total row at the bottom.

However, for the [Count of all (Explicit) Sales] calculated measure in the final column, the ALL function is being used to specify that all implicit filter context that may exist on the 'Fact Sale' table should be removed. The result is the same value of 114,687, because the implicit filtering on [Calendar Year] is being removed using the ALL function. This shows that the query context has been removed by the ALL function, while the upstream filtering that 'Fact Sale'[Quantity] >=10 remains.

Perhaps a more meaningful name for the ALL function might have been REMOVEALLIMPLICITFILTERS, which perhaps is a better indication of the objective of the function. I have used the term *implicit* throughout this chapter. The key point is although it is effective at removing all inherited filters, an explicit filter, as used here to limit the result to rows where the [Quantity] is greater than or equal to 10, is not affected.

The ALL Function

The previous section introduced the powerful ALL function. Let's take a closer look at this function. The syntax is as follows:

```
ALL (  <table> | <column> [, <column> ...]  )
```

The | symbol between the <table> and the first <column> value signifies an "or." This means the first parameter can be a table OR a column. If the first parameter is a table, there is no option for a second parameter. In this case, all implicit filter context is removed from the specified table, which effectively returns all rows from the table.

If a column is used as the first parameter, then additional columns can be used as the second and third parameters as long as all the columns come from the same table.

A more efficient use of the [Count of all (Explicit) Sales] calculated measure from the previous section would be

```
Count of all (Explicit) Sales =
                    CALCULATE(
                        [Count of Sales (10 or more)],
                        ALL('Dimension Date'[Calendar Year])
                        )
```

The change to the calculated measure is that the ALL function is now using a column instead of a table as its parameter. More importantly, the column being passed to the ALL function is the same column being used to drive the implicit filtering. The values produced by this updated version are the same as the version that used the 'Fact Sales' table as a parameter.

In this example, different values are produced for each cell if any other column is used by the ALL function. The query plan generated for both versions of these calculated measures was the same, so swapping a <table> for a <column> here has no effect on performance. In other cases, swapping a <table> for a <column> can provide faster performance.

The ALLEXCEPT Function

Another DAX function that provides the ability to override implicit filters is the ALLEXCEPT function. This is similar to the ALL function in the way it can be used to clear filters, however, with this function, the parameters are used to specify which filters retain implicit filters.

This is the syntax of the ALLEXCEPT function:

```
ALLEXCEPT ( <table>, <column> [, <column> ...] )
```

With ALLEXCEPT, the minimum requirement is to pass both a table and a column. One or more columns can be used as additional parameters. This function provides a handy alternative to the ALL function when used on tables with lots of columns.

If you have a table with 30 columns and a calculation that needs to clear implicit filters on 28 of these, using the ALL function requires 28 parameters to be passed. However, using the ALLEXCEPT function only requires 3 columns to be passed.

Consider the Table 6-1.

Table 6-1. *Dataset to Be Used as Table1 Table*

ColumnA	ColumnB	ColumnC	ColumnD	ColumnE
A	Apple	Australia	Athletics	10
B	Banana	Brazil	Athletics	15
C	Cherry	Canada	Cricket	30

The calculated measures in Listings 6-7 and 6-8 have the same net effect.

Listing 6-7. A Calculated Measure Using the ALL Function with Specific Columns

```
ALL Measure =
          CALCULATE(
              SUM('Table1'[ColumnE]),
              ALL(
                  Table1[ColumnA],
                  Table1[ColumnB],
                  Table1[ColumnC]
                  )
          )
```

Listing 6-8. A Calculated Measure Using the ALLEXCEPT Function with Specific Columns

```
ALLEXEPT Measure =
          CALCULATE(
              SUM('Table1'[ColumnE]),
              ALLEXCEPT(
                  Table1,
                  Table1[ColumnD]
                  )
          )
```

The result of adding both calculated measures to a table visual is shown in Figure 6-12.

ColumnA	ColumnB	ColumnC	ColumnD	ColumnE	ALL Measure	ALLEXEPT Measure
A	Apple	Australia	Athletics	10	25	25
B	Banana	Brazil	Athletics	15	25	25
C	Cherry	Canada	Cricket	30	30	30
Total				55	55	55

Figure 6-12. *Output using code from Listings 6-7 and 6-8*

The two right-hand columns in Figure 6-12 show the output of both versions of the calculated measure. In each case, implicit filters being generated by the first three columns are being cleared leaving only implicit filters from ColumnD to be considered by the SUM function inside CALCULATE.

What this also shows is that you can perform Percentage of Total calculations against subcategories within your data. Using either the ALL or ALLEXCEPT function in this way, you can generate a value showing the sum of values in ColumnE, grouped by ColumnD. You can then use this in a calculation that shows the proportion of ColumnE that is Brazil over every row that has Athletics in ColumnD.

The calculated measure in Listing 6-9 returns a value that represents a percentage of a subcategory. The results are shown in Figure 6-13.

Listing 6-9. A Calculated Measure Incorporating a Calculated Measure from Listing 6-7

```
ColumnD Ratio =
        DIVIDE(
            SUM('Table1'[ColumnE]),
            [ALL Measure],
            0
            )
```

ColumnA	ColumnB	ColumnC	ColumnD	ColumnE	ALL Measure	ColumnD Ratio
A	Apple	Australia	Athletics	10	25	40.00%
B	Banana	Brazil	Athletics	15	25	60.00%
C	Cherry	Canada	Cricket	30	30	100.00%
Total				55	55	100.00%

Figure 6-13. *Output of code from Listing 6-9 when formatted as a percent*

The ALLSELECTED Function

The final function to look at in this section is the ALLSELECTED function. Like ALL and ALLEXCEPT, the purpose of this function is to remove implicit or inherited filtering from the current calculation to allow access to rows of data that would otherwise be unavailable. The difference between ALLSELECTED and the earlier functions is that ALLSELECTED only removes implicit filters that are generated within the same query.

This is especially useful in Power BI where each value on a visual can be influenced by filters from many places. Values can be filtered explicitly through hardcoded DAX or implicitly through other fields in the same visual such as column or row headers (or fields used on an axis). Lastly, filters can be inherited from selections made external to the current visual. It is the distinction between implicit filters generated by filters on fields in the current visual and external filters where ALLSELECTED is most useful.

To get a feel for using all of these functions with WideWorldImportersDW data, consider the Listings 6-10, 6-11, and 6-12.

Listing 6-10. Calculated Measure Using SUM

```
Sum of Quantity = SUM('Fact Sale'[Quantity])
```

Listing 6-11. Calculated Measure Using the ALL Function to Manage Filter Context

```
ALL Quantity =
    CALCULATE(
        SUM('Fact Sale'[Quantity]),
        ALL('Fact Sale')
        )
```

Listing 6-12. Calculated Measure Using the ALLSELECTED to Manage Filter Context

```
ALLSELECTED Quantity =
    CALCULATE(
        SUM('Fact Sale'[Quantity]),
        ALLSELECTED('Fact Sale')
        )
```

When these measures are used in a table visual along with the 'Dimension Date'[Calendar Year] column, you get the result in Figure 6-14.

Calendar Year ▲	Sum of Quantity	ALL Quantity	ALLSELECTED Quantity
2013	2,401,657	8,950,628	8,950,628
2014	2,567,401	8,950,628	8,950,628
2015	2,740,266	8,950,628	8,950,628
2016	1,241,304	8,950,628	8,950,628
Total	**8,950,628**	**8,950,628**	**8,950,628**

Figure 6-14. *Output using code from Listings 6-10, 6-11, and 6-12*

There is no difference between the values in the [ALL Quantity] and [ALLSELECTED Quantity] columns. The assumption here is that no other visuals or external filters are being used to generate this result. Both the ALL and ALLSELECTED functions are removing the implicit filtering being driven from the [Calendar Year] column.

However, when an external filter is added to the report, such as a slicer over a field in a related table, you can see how the values are affected (Figure 6-15).

Calendar Year ▲	Sum of Quantity	ALL Quantity	ALLSELECTED Quantity	Buying Package
2013	1,725,397	8,950,628	6,322,488	☐ Carton
2014	1,829,672	8,950,628	6,322,488	■ Each
2015	1,964,258	8,950,628	6,322,488	☐ N/A
2016	803,161	8,950,628	6,322,488	☐ Packet
Total	**6,322,488**	**8,950,628**	**6,322,488**	

Figure 6-15. *Output of the code from Listings 6-10, 6-11, and 6-12 with filtering from a slicer*

In this case, a slicer has been added using the 'Dimension Stock Item'[Buying Package] field and a selection of Each has been made. The three calculated measures now react differently to this external filter.

The [Quantity] calculated measure now generates a sum over the Quantity field, and it inherits two layers of implied filter context. The value of 1,725,397 for the top row is first influenced by the [Calendar Year] field filtering rows in the 'Fact Sale' table that are relevant to 2013, while the second layer of filtering that comes from the slicer is set to Each. Both layers accumulate to leave the SUM function to only use rows that match both criteria. No filters are overridden for this calculated measure.

The [ALL Quantity] calculated measure uses the ALL function to remove any implied filter context from the calculation. This removes filtering from both the internal layer being driven by the [Calendar Year] field and the external layer of implicit filtering being driven by the slicer. The net result is every cell has the same value including the total. This also happens to be the sum of every 'Fact Sale'[Quantity] column.

The [ALLSELECTED Quantity] calculated measure now shows a different value than the [ALL Quantity] column. This is because the ALLSELECTED function removes the internal layer of filtering being driven by the [Calendar Year] column, but it doesn't remove any filtering coming from the slicer. If a different selection is now made to the slicer, the [ALLSELECTED Quantity] measure updates with new values, while the [ALL Quantity] measure remains unchanged.

The three calculated measures in this example can be used as the basis for additional measures that highlight data in interesting ways. All are potentially useful in your report. Hopefully these examples show the intent and approach of the different filter functions. It's possible to target a very fine grain of row selection using these functions to provide extensive flexibility.

Running Totals

The ability to remove implicit filters from calculations is also useful when used to generate running totals. To create a calculated measure that shows the running total of the 'Fact Sale'[Quantity] column, start with a core DAX expression such as this:

```
SUM('Fact Sale'[Quantity])
```

The challenge is that when you use this expression in a visual, the calculation only considers values in the 'Fact Sale'[Quantity] column that are implicitly filtered by other fields. If the 'Dimension Date'[Calendar Year] field is also used in the query, the result produced for this calculation correctly shows the SUM for each specific calendar year but does not have access to data relating to other years to provide a running total effect. This is where the ALL, ALLEXCEPT, and ALLSELECTED functions can help.

To see a demonstration of running totals using WideWorldImportersDW data, consider the following calculated measures in Listings 6-13 and 6-14.

Listing 6-13. A Calculated Measure Using SUM

```
Sum of Quantity = SUM('Fact Sale'[Quantity])
```

Listing 6-14. A Calculated Measure Using the FILTER Function to Generate a Running Total

```
Running Total =
    CALCULATE(
        SUM('Fact Sale'[Quantity]),
        FILTER(
            ALL('Dimension Date'[Calendar Year]),
            MAX('Dimension Date'[Calendar Year]) >= 'Dimension
            Date'[Calendar Year])
            )
```

When these calculated columns are added to a visual that also uses the 'Dimension Date'[Calendar Year] column, the results in Figure 6-16 are produced.

Calendar Year	Sum of Quantity	Running Total
2013	2,401,657	2,401,657
2014	2,567,401	4,969,058
2015	2,740,266	7,709,324
2016	1,241,304	8,950,628
Total	**8,950,628**	**8,950,628**

Figure 6-16. *Output of the code from Listings 6-13 and 6-14*

The values in the [Sum of Quantity] column all show the result of the calculation being run with the relevant implicit filter being driven from the [Calendar Year] column. The bottom value of 8,950,628 in this column has no implicit filter context, so it calculates over every row in the 'Fact Sale' table.

For the [Running Total] calculation, the SUM function is wrapped inside a CALCULATE function that allows additional filtering instructions to be defined, in this case, the targeted removal of some implicit filtering. For the result of 4,969,058 for 2014, the SUM function needs to have access to rows that relate to 2013.

The first parameter of the FILTER function uses the ALL function to remove any implicit filtering from the [Calendar Year] column. This has the effect of exposing every row from the 'Fact Sale' table to the SUM function. If you stopped here, you would simply see the value of 8,950,628 repeated in every cell.

The second parameter of the FILTER function specifies MAX ('Dimension Date'[Calendar Year]) >= 'Dimension Date'[Calendar Year]), which defines a filtering rule that now reduces rows from the 'Fact Sale' table to only those that mean this Boolean condition. For the first row in the table visual, this means only rows that related to 2013. For the second row, this includes all rows that relate to 2013 as well as 2014. This is the magic that provides the effect of a running total.

Each cell in the [Running Total] column is an independent calculation that does not rely on, or use, the output from any previous calculation in the visual. The calculations can run in any order and will only consider rows that survive the multiple layers of filtering being applied.

DAX can perform these calculations very quickly due to the in-memory, column-based nature of the xVelocity engine, so the extra flexibility provided through the independent nature of cell-based calculations generally outweighs any downsides of read duplication. It is good to keep this in mind if you are building running totals over many calculations. In this example, the [Running Total] calculation executes five times.

If the 'Dimension Date'[Date] column was used in place of 'Dimension Date'[Calendar Year] for both the calculated measure and the visual, there would be 1,462 executions of the calculated measure (one for each row at the [Date] level and then one for the total). Any values in rows related to January 1, 2013, would be used by every one of the 1,462 executions.

ALLSELECTED on Running Total

The [Running Total] calculated measure shown previously in Listing 6-14 used the ALL function to remove any implicit filters being driven by the 'Dimension Date'[Calendar Year] column. When employed in this manner, other slicers and filters used external to the visual that are based on columns other than 'Dimension Date'[Calendar Year] filter through to the SUM calculation. This means they give a running total from the start of the first record in the [Calendar Year] column, but the rows also have an additional filter applied that is relevant to the selection of the slicer.

The exception to this is if a slicer or external filter is based on the 'Dimension Date'[Calendar Year] column. In this case, the value produced if the slicer is filtered to 2015 is still 7,709,324. All rows that relate to 2013 and 2014 are still considered by the SUM.

If the calculation is changed to use the ALLSELECTED function in place of the ALL function, any slicer or external filter based on 'Dimension Date'[Calendar Year] now only considers rows that relate to selections in the slicer.

This can be demonstrated using the three calculated measures in Listings 6-15, 6-16, and 6-17.

Listing 6-15. A Calculated Measure Using SUM

```
Sum of Quantity = SUM('Fact Sale'[Quantity])
```

Listing 6-16. Running a Total Calculated Measure Using the ALL Function to Manage the Filter Context

```
Running Total ALL =
    CALCULATE(
        SUM('Fact Sale'[Quantity]),
        FILTER(
            ALL('Dimension Date'[Calendar Year]),
            MAX('Dimension Date'[Calendar Year]) >= 'Dimension
            Date'[Calendar Year])
            )
```

Listing 6-17. Running a Total Calculated Measure Using the ALLSELECTED Function to Manage Filter Context

```
Running Total ALLSELECTED =
    CALCULATE(
        SUM('Fact Sale'[Quantity]),
        FILTER(
            ALLSELECTED('Dimension Date'[Calendar Year]),
            MAX('Dimension Date'[Calendar Year]) >= 'Dimension
            Date'[Calendar Year])
            )
```

Figure 6-17 shows the result when a selection has been made to a slicer using the 'Dimension Date'[Calendar Year] column. In this example, two selections are made to the slicer for 2015 and 2016.

Calendar Year ▾	Sum of Quantity	Running Total ALL	Running Total ALLSELECTED	Calendar Year
				☐ 2013
2016	1,241,304	8,950,628	3,981,570	☐ 2014
2015	2,740,266	7,709,324	2,740,266	☑ 2015
Total	**3,981,570**	**8,950,628**	**3,981,570**	☑ 2016

Figure 6-17. *Output of code from Listings 6-15, 6-16, and 6-17 with filtering from slicer using two selections*

The difference between the two running total calculated measures is the way each reacts to the slicer. The version that uses ALL removes all filtering over the 'Dimension Date'[Calendar Year] column, regardless of where the filter originated, whereas the version using ALLSELECTED retains the filtering context from the slicer. This means the calculation using ALLSELECTED cannot access rows from 'Fact Sale' that relate to 2013 or 2014, so the running total is started from 2015 rather than from 2013.

Either approach may suit your reporting requirements. Sometimes you need to use one approach, other times you need the other. This example hopefully demonstrates how you can use the different filter functions to achieve what you need.

Resetting Running Totals

A slight variation on the running total measure is to provide a running total over a value for a subcategory that resets between categories. For instance, this can be demonstrated using the 'Dimension Date' table, where the category is Calendar Year and the subcategory is Calendar Month.

The following calculated column is first added to the 'Dimension Date' table:

```
Calendar Month = STARTOFMONTH('Dimension Date'[Date])
```

This generates a new column that carries a value that shows the first day of every relevant month.

A modified version of the [Running Total ALL] calculated measure with differences highlighted is shown in Listing 6-18.

Listing 6-18. Running Total Calculated Measure That Resets Each Year

```
Running Total ALL =
    CALCULATE(
        SUM('Fact Sale'[Quantity]),
        FILTER(
            ALL(
                'Dimension Date'[Calendar Month],
                'Dimension Date'[Calendar Year]
                ),
            MAX('Dimension Date'[Calendar Month]) >= 'Dimension
            Date'[Calendar Month] &&
            MAX('Dimension Date'[Calendar Year]) = 'Dimension
            Date'[Calendar Year])
            )
```

When used in a Matrix visual with the 'Dimension Date'[Calendar Year] and the new 'Dimension Date'[Calendar Month] columns, the calculated measure produces the result in Figure 6-18.

Calendar Year	Sum of Quantity	Running Total ALL
2013	**2,401,657**	**2,401,657**
January 2013	193,271	193,271
February 2013	142,120	335,391
March 2013	207,486	542,877
April 2013	212,995	755,872
May 2013	230,725	986,597
June 2013	213,468	1,200,065
July 2013	232,599	1,432,664
August 2013	192,199	1,624,863
September 2013	190,567	1,815,430
October 2013	198,476	2,013,906
November 2013	194,290	2,208,196
December 2013	193,461	2,401,657
2014	**2,567,401**	**2,567,401**
January 2014	216,337	216,337
February 2014	182,103	398,440
March 2014	196,451	594,891
April 2014	209,020	803,911
May 2014	239,381	1,043,292

Figure 6-18. *Output using code from Listing 6-18. Running totals are reset for 2014*

In this calculation, the FILTER function uses more sophisticated rules to determine which filters should be removed or applied to decide which rows can be considered by the SUM function.

The ALL function removes any implied filtering from the [Calendar Year] and [Calendar Month] columns. At this point, the SUM calculation returns the value of 8,950,628 for every cell in the [Running Total ALL] column.

However, additional filtering is being applied by rules in the FILTER function. Two conditions need to be met according to these rules.

The first condition is that rows in the 'Fact Sale' table related to each 'Dimension Date'[Calendar Month] must be the same, or lower (earlier) than the Calendar Month from the row header. This is the condition that provides the running total effect.

This is a confusing notation. There appear to be two references to the same 'Dimension Date'[Calendar Month] object. These are not the same objects. One refers to the version of 'Dimension Date' in the query context, while the other refers to the version of 'Dimension Date' used in the filter context.

The 'Dimension Date'[Calendar Month] wrapped in the MAX function on the left-hand side of the >= operator is referring to the query context. The MAX function guarantees just one value to be used in the comparison.

The 'Dimension Date'[Calendar Month] on the right-hand side of the >= operator refers to the filter context. It is these records, including downstream rows in 'Fact Sale' that are used and passed to the SUM function.

The other criteria in the filter expression is to specify only rows from 'Fact Sale' that have a relationship to 'Dimension Date'[Calendar Year] that matches the value from the Calendar Year row header. This is the part of the code that resets the running total.

This requirement could also be satisfied using the following calculated measure:

```
Running Total YTD =
    TOTALYTD(
            [Sum of Quantity],
            'Dimension Date'[Date]
            )
```

The TOTALYTD function provides a running total for each Calendar Month as well as resetting each year. Although this is simpler, it is only designed to work with date-based columns and it provides less flexibility around your running total. The [Running Total ALL] version uses a pattern that you can apply to different types of data, with more control over the rules around when to reset the running total (if at all).

Running Totals as a Rank (Calculated Column)

An alternative use of the running total pattern is to provide a ranking over a set of data with the output being a column rather than a measure. This calculated column provides a version that establishes a ranking from 1 to 403 for customers based on overall sales (Listing 6-19). The calculation relies on a one-to-many relationship existing between the 'Dimension Customer' and 'Fact Sale' tables.

Listing 6-19. A Calculated Column Using RANKX Over a Relationship

```
Customer Rank Column =
    RANKX(
        'Dimension Customer',
        CALCULATE(
            SUM('Fact Sale'[Total Excluding Tax])
            )
        )
```

An interesting feature to point out about this DAX statement is the use of CALCULATE. This is required to convert the current row context to a filter context to allow the SUM function to correctly generate the correct value for each row using information across the relationship. If the CALCULATE statement is removed, the SUM function simply returns the same value over and over. The value returned is the sum of every row in the 'Fact Sale' table.

This calculation only computes when data is being refreshed into the model because it is a calculated column. Filters and slicers have no effect on the number returned.

To help understand the approach of the DAX formula, take a look at the T-SQL equivalent (Listing 6-20).

Listing 6-20. The T-SQL Equivalent of the DAX Code in Listing 6-19

```
WITH Ranking (
     [Customer Key],
     [Total Excluding Tax]
     )
     AS
     (
     SELECT
             D.[Customer Key],
             SUM(F.[Total Excluding Tax]) AS [Total Excluding Tax]
     FROM Dimension.Customer AS D
             INNER JOIN Fact.Sale AS F
                 ON F.[Customer Key] = D.[Customer Key]
     GROUP BY
             D.[Customer Key]
             )
```

```
SELECT
        L.[Customer Key],
        COUNT(*) AS [Final Ranking]
          FROM Ranking AS L
        INNER JOIN Ranking AS R
              ON L.[Total Excluding Tax] <= R.[Total Excluding Tax]
    GROUP BY
        L.[Customer Key]
    ORDER BY
        COUNT(*)
```

The final select statement performs a self-join back to the predicate on the INNER JOIN filtering out rows with a higher value for [Total Excluding Tax]. You can obtain a ranking by counting the number or rows after the INNER JOIN. You can achieve this in other ways using T-SQL, but this version is useful to help you understand the approach taken by the DAX formula.

The FILTER function takes a similar approach to a T-SQL self-join, which you can tailor to suit different requirements by using different filtering rules.

Period Comparison

A common requirement is to compare a value for one period to a previous period. This allows further calculations to show variation percentages, among other things. One example is showing a value for a measure for a specific month next to a value of the same measure for the previous month (or the same month from a previous year).

Consider the example of comparing to a previous month when you take a look at the four calculated measures in Listings 6-21–6-24.

Listing 6-21. A Period Comparison Calculated Measure Using the ALL Function to Manage Filter Context

```
Using Filter =
    CALCULATE(
        [Sum of Quantity],
        FILTER(
            ALL('Dimension Date'[Calendar Month]),
```

```
      MAX('Dimension Date'[Calendar Month]) =
      EDATE('Dimension Date'[Calendar Month],1)
      )
   )
```

Listing 6-22. A Period Comparison Calculated Measure Using PARALLELPERIOD

```
Using Parallel Period =
    CALCULATE(
        [Sum of Quantity],
        PARALLELPERIOD('Dimension Date'[Date],-1,MONTH)
        )
```

Listing 6-23. A Period Comparison Calculated Measure Using DATEADD

```
Using Date Add =
    CALCULATE(
        [Sum of Quantity],
        DATEADD('Dimension Date'[Date],-1,MONTH)
        )
```

Listing 6-24. A Period Comparison Calculated Measure Using
PREVIOUSMONTH

```
Using Previous Month =
    CALCULATE(
        [Sum of Quantity],
        PREVIOUSMONTH('Dimension Date'[Date])
        )
```

All four calculated measures produce the same result, which you can see in Figure 6-19 as the columns on the right-hand side.

Calendar Month	Sum of Quantity	Using Filter	Using Parallel Period	Using Previous Month	Using Date Add
January 2013	46,176				
February 2013	33,908	46,176	46,176	46,176	46,176
March 2013	53,000	33,908	33,908	33,908	33,908
April 2013	54,054	53,000	53,000	53,000	53,000
May 2013	55,990	54,054	54,054	54,054	54,054
June 2013	52,822	55,990	55,990	55,990	55,990
July 2013	57,462	52,822	52,822	52,822	52,822
August 2013	47,818	57,462	57,462	57,462	57,462
September 2013	48,520	47,818	47,818	47,818	47,818
October 2013	51,166	48,520	48,520	48,520	48,520
Total	**2,320,536**		**2,320,536**		**2,320,536**

Figure 6-19. *Output of code from Listings 6-21–6-24*

For the February 2013 row, all four calculated measures return the value of 46,176, which happens to be the value of [Sum of Quantity] for January 2013. Aside from the fact that two of the measures also provide a value in the [Total] column, they all react to external slicers and filters.

The [Using Filter] measure, is the most complex to write; however, it allows the most flexibility for customizing the filtering rules. It is easy to add and apply additional layers of explicit filtering in this instance. You can also configured the [Using Filter] measure to always look at a specific month rather than a relative month.

The [Using Parallel Period] and [Using Date Add] versions take advantage of functions that don't require the ALL or ALLSELECTED functions to access out-of-context data. These functions also provide the ability to customize to suit other time periods. Perhaps the requirement is to always compare the current month with one that is always two months prior, or twelve months prior. Both the PARALLELPERIOD and DATEADD functions allow QUARTER and YEAR as options to return. You can also use them with positive numbers and negative numbers for the interval.

Finally, the [Using Previous Month] measure provides the simplest format of the four measures but the least flexibility.

In this case, all four calculated measures are evaluating an expression over a column in the 'Fact Sale' table, but they use a column in the 'Dimension Date' table to determine the date period. These use the rule determined by the active relationship between these tables, which in this case is based on the 'Fact Sale'[Invoice Date key]. If the calculation should be grouped using a different column in 'Fact Sale', then you can add USERELATIONSHIP as additional parameter to the CALCULATE function. This allows a visual to show values that represent measures by delivery date if that is what you require.

The 'Dimension Date'[Date] column also features in each of these calculated measures. This column exists in a table where contiguous dates are guaranteed. Some of the time intelligence–based measures rely on this to work. If a date column from 'Fact Sale' is used, unexpected results may occur that may not initially be obvious.

I analyze a breakdown of the individual performance of these measures in Chapter 8.

Calculated Columns and the EARLIER Function

A handy function to use when you are writing calculated columns is the EARLIER function. This function provides a logical equivalent of a T-SQL self-join, which allows calculations to access data from rows other than the current row. This can include data from rows in related tables. The name of this function doesn't clearly explain how it works; perhaps an alternative name might be more useful.

To see how this function can be used, add a calculated column to the 'Fact Sale' table. This column should carry values that show, for each transaction, the date of the previous transaction for that customer. You can then use this calculation as the basis for determining the number of days since last purchase and it can then be averaged and plotted over time in a way that helps you understand changes to the frequency of visits by customers.

The DAX calculated column is shown in Listing 6-25.

Listing 6-25. A Calculated Column Using EARLIER

```
Last Purchase Date =
    CALCULATE(
        LASTDATE(' Fact Sale'[Invoice Date Key]),
        FILTER(
            'Fact Sale',
            'Fact Sale'[Customer Key] = EARLIER('Fact Sale'[Customer Key])
            && 'Fact Sale'[Invoice Date Key] < EARLIER('Fact Sale'[Invoice
            Date Key])
        )
    )
```

When this calculation is added to the 'Fact Sale' table, a new column shows for each row the last time that specific customer was invoiced.

The T-SQL equivalent is shown in Listing 6-26.

Listing 6-26. The T-SQL Equivalent of the DAX code in Listing 6-25

```
SELECT
      [Current].[Sale Key],
      [Current].[Customer Key],
      [Current].[Invoice Date Key],
      MAX([EARLIER].[Invoice Date Key]) AS [Last Purchase Date]
FROM Fact.Sale AS [Current]

      LEFT JOIN FACT.Sale AS [EARLIER]
            ON  [Earlier].[Customer Key] = [Current].[Customer Key]
            AND [Earlier].[Invoice Date Key] < [Current].[Invoice Date
            Key]

WHERE
      [Current].[Customer Key] = 10

GROUP BY
      [Current].[Sale Key],
      [Current].[Customer Key],
      [Current].[Invoice Date Key]
```

In order for each row in 'Fact Sale' to find the correct value to use as the last purchase date, it needs access to information that isn't contained in its row. It needs a way to look at other rows from the table. The EARLIER function allows a form of logical self-join that, when used in the FILTER function, allows the calculation to find the relevant rows.

In this case, a single row in the 'Fact Sale' table first finds every row from the self-join table that is a match on [Customer Key]. This includes rows that have an [Invoice Date Key] after the value from the current row. Then the additional filter requirement that limits rows from the outer [Invoice Date Key] so they have a smaller value than the [Invoice Date Key] of the current row is applied.

This can still produce multiple values so the LASTDATE function is used in the expression to pick a single date that happens to be the oldest of the remaining dates. You could also use the MAX function in place of LASTDATE in this case.

This calculated column produces a value that is relevant for every row in the table, and because it is a calculated column rather than a calculated measure, it cannot be changed using report slicers or filters.

The name of the EARLIER function implies that it may provide functionality that looks for a previous row based on ordering. This is not the case, and perhaps a name like OUTERTABLE would be more meaningful.

Variables provide an alternative notation for the EARLIER function. The calculated column in Listing 6-27 does the same job using VAR instead of EARLIER.

Listing 6-27. A Calculated Column Using Variables in Place of EARLIER per Listing 6-25

```
Last Purchase Date =
VAR CustomerKey = 'Fact Sale'[Customer Key]
VAR InvoiceDateKey = 'Fact Sale'[Invoice Date Key]
RETURN
    CALCULATE(
        LASTDATE('Fact Sale'[Invoice Date Key]),
        FILTER(
            'Fact Sale',
            'Fact Sale'[Customer Key] = CustomerKey
            && 'Fact Sale'[Invoice Date Key] < InvoiceDateKey
        )
    )
```

The differences between this and the original version using EARLIER have been highlighted and they make the CALCULATE code a little easier to read.

Filters and Calculated Tables

If you are using a version of DAX that allows the use of calculated tables, a simple use of the FILTER function can be to create copies of tables filtered to certain conditions. This might be useful for debugging, or if you need to solve performance challenges in your data model.

If your dataset is large and you have some calculations that you know will only ever run over a subset of the same data, then you can use DAX to create calculated tables from existing tables with explicit filters defined.

An example of this might be if you need to create a smaller version of the 'Dimension Customer' table that only includes the top ten customers defined by the 'Customer Rank' calculation at Listing 6-19.

You might find this useful for scenarios in which you need a mixture of behavior for customer-based slicers in your report page.

The DAX for the calculated table using the FILTER function is shown in Listing 6-28.

Listing 6-28. A Calculated Table Using FILTER

```
Top 10 Customers =
    FILTER(
        'Dimension Customer',
        'Dimension Customer'[Customer Rank] <= 10
        )
```

If a new customer generates enough activity to jump into the top ten based on the rules defined in the calculated column, this calculated table will dynamically reflect that.

CHAPTER 7

Dates

Most data models involve an element of date or time. I can't remember the last time I worked on a data model that didn't involve date values anywhere in the data.

Business data invariably involves tables of transactions that include a point-in-time value. Sometimes a row of data might carry multiple DateTime fields to represent different actions relevant to that transaction. In the case of a customer order, there may be values to show the point in time of the original order, the pick-up date, the delivery date, and the invoice date. Downstream reporting requirements may need to group data by any of these DateTime values to provide a perspective relevant to different parts of the business.

Sensor log data may include a field that identifies the sensor, another that shows the point in time of the reading (possibly down to a microsecond), and the value for the reading itself.

Other types of data inevitably have at least one component that records a DateTime value that represents a point in time that is relevant for the other data in the row or batch.

For reporting and analytics, data models often need to organize raw data using the following three types of questions:

- *What has happened:* For example, "Where have we come from?"

- *The current state:* For example, "Where are we now?"

- *The future:* For example, "Where are we going?"

To answer each of the questions will require the ability to organize data by date or time. The first helps show values for facts and events that have taken place historically. This might illustrate patterns for a metric that ebb and flow historically. For instance, recent data can be trending up or down. Or cyclic or seasonal trends might appear and be useful when you are grouping, sorting, and plotting data over time. Being able to group and sort historic data into time buckets makes the job of understanding what has happened much easier.

© Philip Seamark 2018
P. Seamark, *Beginning DAX with Power BI*, https://doi.org/10.1007/978-1-4842-3477-8_7

It can be useful to show the current state of data in short-term planning. To know how much money is currently available in an account, or how many items are currently in stock, or the current location of a vehicle or item can provide useful insights that help you when you are developing short-term plans.

Finally, being able to forecast likely values for upcoming points in time satisfies a separate set of reporting requirements. Using forecast data to help accurately size and optimize future staff and stock levels can save an organization valuable time and resources. This information can lead to improvements in overall efficiency.

All three requirements depend on date and time data in some shape or form. Organizing data into buckets and groups that are sorted chronologically makes the task of creating useful reports much easier.

Date

In DAX, the term *Date* often refers to individual days or groupings of days where there is no detail of hours, minutes, or seconds. Even if the underlying data carries a value that represents hours, minutes, or seconds, Date likely indicates the value without hours, minutes, or seconds. Dates can then be organized into many group types. Obvious groups are calendar year or month, or fiscal year or month. Other useful groupings can include weeks, or days of the week, month, or year. There are many ways to group Dates. Some are common to diverse types of data, whereas others can be highly unique to your organization.

Plenty of meaningful reports can be generated using data where the lowest granularty is Date. Even if the raw data happens to carry information regarding the point in time to the second a transaction happened, stripping out hours, minutes, and seconds to show the number of sales per day, month, or year, is often more useful than reports that represent the finer level of detail.

Time

In data modelling, Time is often used to refer to a point in time that includes a value for the hour, minute, and second (sometimes millisecond). This may or may not have a value for year, month, and day. Sometimes a Time value can be perfectly meaningful with a value such as 10:37 am and it does not matter which day it belongs to. This value can be broken up into groupings such as 10 am, or Morning (or Late Morning), with data aggregated and used to show or compare.

Date/Calendar Tables

A useful data structure for most data models is to include a table that carries a set of rows that covers a range of individual days. This table is often referred to as a Date or Calendar table. Sometimes these tables cover a very wide range of time. In other variations, the earliest date in this table may only be around the time of the earliest day found anywhere in your data model. These tables can also have rows that represent days a long way into the future.

A key characteristic of a Date table is that each day is represented by a single row. There should be no duplicate rows for any individual day, and more importantly, there should be no gaps in the data. So, every day that occurs in between the dates that happen to be the oldest and newest values in the table should be represented by a single row in the table.

Time tables have similar characteristics to Date tables. First, they have a row to represent the lowest level of granularty, starting with the lowest value and ending with the highest value. The decision of what granularty should be the lowest for a Time table depends on your reporting requirements. The lowest level of granularty for a Time table can be hours or minutes (or lower). In the case of minutes, the table should have 1,440 rows that individually represent every minute between midnight and 11:59 pm. If the lowest granularty happens to be hour, only 24 rows are required.

When creating a relationship between a Date (or Time) table and other tables in your data model, place the Date table on the one side of the relationship. The column involved in the relationship needs to be unique and can be any datatype. Ideally the column in the related table is the same datatype and only has values that can be exactly matched to values from the column in your Date table. It's perfectly fine to have rows in your Date table that have no matching rows in the related table, but you have a problem if the mismatch is the other way around.

Note   If the column used in a relationship from your Date table does not contain hours, minutes, or seconds, whereas the column from the related table does, many rows may not match. If this is the case, visuals that use fields from the Date table on an axis will report lower-than-expected values. No error will be generated, so this is potentially easy to miss.

For Date tables, I typically like to use the column that uses the Date datatype, although integer values such as 20190101 work just as well. The main thing is to make sure that the values in each column involved in the relationship use the same data format.

Date tables can have relationships to multiple tables at the same time. This allows selections made using a slicer (or filter) that uses the Date table to propagate through to all related tables.

It's possible to create multiple relationships between a Date table and another table in the model. As mentioned earlier, a sales table may have columns that represent various stages of an order. The order date, delivery date, and invoice date may all have different values on the same row. Figure 7-1 shows two relationships between 'Dimension Date' and 'Fact Sale.' The solid line between the two tables represents the active relationship. Calculations using data from tables will use this relationship by default. If a calculation needs to use rules from an inactive relationship, it can use the USERELATIONSHIP function as part of the calculation.

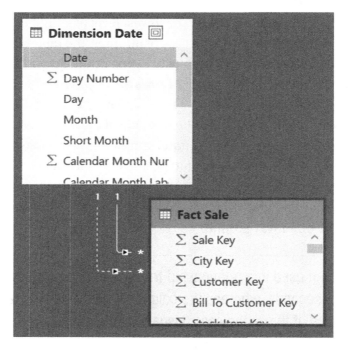

Figure 7-1. *Relationships between 'Dimension Date' and 'Fact Sale'*

In this diagram, the active relationship between 'Dimension Date' and 'Fact Sale' is based on the [Invoice Date Key], whereas the inactive relationship uses the [Delivery Date Key].

The calculations shown in Listings 7-1 and 7-2 show how relationships can be used. The [Count of Invoice Rows] calculated measure only contains an expression about data from 'Fact Sale'. When this measure is used in a visual that has fields from 'Dimension Date', query context will automatically filter groups of rows from 'Fact Sale' according to [Invoice Date Key].

The [Count of Delivery Rows] measure uses the USERELATIONSHIP to override the default behavior in order to generate a value that can be used to plot the number of rows where a delivery was made for a given group of 'Dimension Date' values.

Listing 7-1. Calculated Measure Using COUNTROWS

```
Count of Invoice Rows =
    COUNTROWS('Fact Sale')
```

Listing 7-2. Calculated Measure Using USERELATIONSHIP over Inactive Relationship

```
Count of Delivery Rows =
    CALCULATE(
        COUNTROWS('Fact Sale'),
        USERELATIONSHIP(
                'Dimension Date'[Date],
                'Fact Sale'[Delivery Date Key]
                )
    )
```

Figure 7-2 shows both these calculated measures over the first six calendar months of the WideWorldImportersDW dataset.

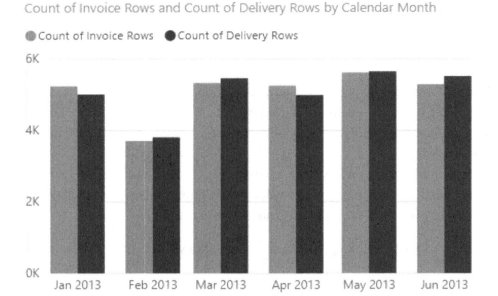

Figure 7-2. *A clustered column chart visual using code from Listings 7-1 and 7-2*

Multiple Date tables are perfectly normal in a data model and allow you to use multiple date ranges in a report. Some calculations can use selections made using a slicer with one Date table, whereas other calculations can produce values that consider selections made using a slicer over another Date table. These are not common, but you can use them to solve trickier reporting scenarios.

The bare minimum requirement of a Date table is having a single column that represents a day. Usually additional columns in the Date table allow for the grouping and ordering of days. Obvious groupings are Year, Quarter, and Month. Some Date tables have dozens of additional columns to suit various date groupings.

Later in this chapter I show you an example of how you can generate a Date table using DAX functions without using external data. Often an organization has an external dataset that you can import into your data model that already contains the rows and columns that will be useful. The query editor also provides you with the ability to generate data for a Date table that you can customized to suit your purposes.

Automatic Date Tables

There is a feature in Power BI Desktop that automatically creates hidden Date tables for your model. If this is enabled, Power BI creates a hidden Date table for every column in every table that uses a Date/DateTime datatype (and is not already involved in one side of a relationship). If your model has five Date/DateTime columns, Power BI Desktop adds five hidden Date tables to the model. The intention of these hidden tables is to help provide an automatic layer of time intelligence to your model.

If you import a table of data to your model and one of the columns in the table uses a Date datatype, Power BI adds a table with seven columns (Table 7-1) that is populated with enough days to cover the data in your imported table, rounded out to the start and end of each year.

Table 7-1. *The Structure and Sample Data from the Power BI Autocreated Date Table*

Date	Year	MonthNo	Month	QuarterNo	Quarter	Day
2019-01-01	2019	1	January	1	Qtr 1	1
2019-01-03	2019	1	January	1	Qtr 1	2
2019-01-03	2019	1	January	1	Qtr 1	3
...	...	...	...	...	...	...

The structure of this table can be a good starting template for any Date table you design. The first column is unique and used in the relationship. The other columns simply provide grouping values. In Power BI, by default, if the Date field from your table is dropped onto a visual, you have the option of using a hierarchy based on this automatic table or simply using the values from your actual table.

This table quickly provides you with the ability to slice and dice by year, quarter, month, or day of month.

If you have a good handle on using Date tables and have a quality dataset of your own to use, you can turn this feature off in the options and settings.

Quick Measures

In 2017, Power BI introduced a feature called Quick Measures. This feature allows report authors to quickly create calculated measures using a dialog box instead of writing DAX by hand. These measures cover a variety of common scenarios such as year-to-date totals and rolling averages. However, at the time of this writing, Quick Measures only uses data from an automatically provided Date table and cannot be used with columns from your own Date table. Once the Quick Measure has been created and you have the pattern for the calculation, you can then modify and tweak it to use other Date tables if you need to.

If you examine the DAX code for a calculated measure created by the Quick Measures feature, you may notice a slightly different syntax is used to reference the column from the automatic Date table. The core expression in a Year-To-Date calculated measure created as a Quick Measures might look like this:

```
TOTALYTD(
        SUM('Table1'[Quantity]),
        'Table1'[My Date Col].[Date]
        )
```

The highlighted part of the preceding code shows a special syntax that only works with automatic Date tables provided by Power BI. There are three parts to this syntax. The first is 'Table1', which is a reference to the table being used. The second is [My Date Col], which is a reference to the column to be used from Table1. The third is the [Date] column reference. This is the name of the column from the automatic Date table. Values from this column will be used by the SUM expression, rather than values from the [My Date Col] column, which might have gaps or other date-based issues.

Sorting by Columns

The structure of the automatic Date table highlights another frequent problem you are apt to encounter when working with Date tables. This has to do with the default sort order of text-based columns. By default, Power BI sorts text-based columns alphabetically. The automatic table has two text-based columns: [Month] and [Quarter]. If either of these columns are used in a visual, the values are sorted alphabetically. In the case of the [Month] column, values like April, August, and December are at the start, while November, October, and September are at the end.

This is a list of the calendar months using the default sort ordering:

```
April
August
December
February
January
July
June
March
May
Novenber
October
September
```

It's highly unlikely that you will ever need to deliberately organize data in this way, however. To override this behavior, use the Sort by Column feature, which allows a column to be sorted using another column. Typically, the nominated column is numeric, but it doesn't have to be. A common pattern in a Date table is that for every text-based column in the table, a numeric equivalent column exists, purely for sorting the text column. The numeric column used for sorting can be used by multiple columns. For instance, you could use [MonthNo] to sort [Long Month Description] as well as [Sort Month Description]. You could also hide the [MonthNo] column to help tidy the model.

The Sort by Column feature is not limited to sorting dates; you can also use it to arrange values like Low, Medium, and High in an order that makes sense for your visuals. Not only does it sort the order of values used in visuals, it also controls the order they appear in legends and filters/slicers. You can also use it to control the order in which categories might appear in the bars of a stacked bar chart.

It's a useful feature, but it requires values in the [Sort By] column to have a many-to-one match with values in the column being sorted. For every row with a value of [January] in the [Month] column, only one distinct value can appear in any of the values used in the [Sort By] column. In this case, every value for "January" would be 1.

Including the Year When Using Month

When planning a Date table, consider including information about the year in any month label or identifier, particularly when the dataset covers more than a year. Rather than converting a date of January 1, 2019, into a value of [January], use [January 2019]. If you don't, calculations may combine data associated with this date with January data from other years.

A good format to use for the numeric version of a month is YYYYMM. This keeps data from the same months, but different years, apart, but it also provides a good column to sort by. Using two uppercase M characters adds a leading zero for months one through nine.

Time Intelligence

A term used regularly when building data models in DAX is *time intelligence*. This essentially refers to the concept of combining calculations/measures with the element of time. It is the ability to not only group and organize data into periods of time, but also to understand how each period sits in relation to any other grouping chronologically. Reporting requirements, such as showing comparisons between related time periods or building periods to date measures, are commonplace when you're designing a data model that is intended to be used for reporting and analytics.

Data grouped into categories such as Year, Month, and Day share similar characteristics to non-time-bound groupings of data. Calculations, such as count, sum, average, and so on, are just as meaningful to time-based groupings as they are to non-time-based groups. However, a key difference between time-based groups and non-time-based (such as Size, Color, and Weight), is that calculations using time-based groups often need to reflect and show how data in your model sits in relation to other data in your model with respect to time.

When you look at this in more detail, counts, sums, and other calculations that are sliced by the product colors Red, Green, Yellow, and Blue, often have no strong relationship with each other. You might display them in any order on a visual and just as happily compare values between Red and Blue as you would Green and Yellow.

With dates however, it is more likely that you'll need to keep categories for Jan 2019, Feb 2019, Mar 2019, and Apr 2019 in sequential order to be able to compare them. Calculations that produce values for Feb 2019 data are more likely to be compared with calculations based on the previous or preceding months. It is still possible to organize DateTime categories in any order, as you might do with the example of using categories based on color, but this is not likely to be a common requirement.

An additional element of time-based reporting is that it can be just as important to show periods where you have no data as it can be to show time periods with data. When you're grouping by color using data that has no values for red, omitting a placeholder for red on the axis of a visual can be quite normal. However, if you have no data for a day, or a date period, you may still want to represent this fact by making sure there is a placeholder for the period on the axis.

DAX provides a series of time-intelligence functions that are aware of chronological relationships and how they relate to each other. You can add these functions to your calculations that take care of common time-based tasks through the minimal use of DAX.

Year to Date

If you need to show the cumulative value of a measure over time since the start of each calendar year, there are several ways you can write this type of calculation. One approach is to create a calculated measure that uses CALCULATE and ALL or ALLSELECTED to clear implicit filters; this allows SUM to access data outside the current query context in order to produce a value since the start of the year.

Another option is to use a calculation to meet this requirement that you can build using a time-intelligence function. With the TOTALYTD function, you can add a calculated measure (Listing 7-3) to your data model that represents a cumulative value of your expression over a calendar year. Figure 7-3 shows the output of this calculation.

Listing 7-3. Year-to-Date Calculated Measure Using TOTALYTD

```
My Year to Date (Date Table) =
    TOTALYTD (
        SUM('Fact Sale'[Quantity]),
        'Dimension Date'[Date]
        )
```

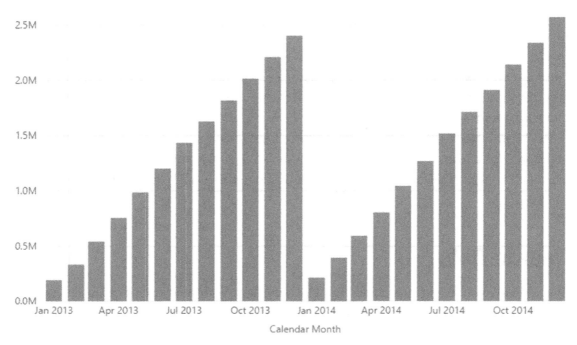

My Year to Date (Date Table) by Calendar Month

Figure 7-3. *Output of the code from Listing 7-3 resetting each calendar year*

The chart in Figure 7-3 shows a bar for each calendar month over a period of two years. Each calculation includes a value that is a sum of the current month and all months since the start of the calendar year. The cumulative values reset each year.

The function requires only two parameters. The first is the expression to be used for a calculation. The SUM function in this calculation uses data from the 'Fact Sale'[Quantity] column. You could also use another calculated measure in place of a DAX expression.

The second parameter needs to be a reference to a date column.

The example doesn't use a date column from the same table as the column passed to the SUM function; instead, it uses a value from a table that conforms to the rules of a classic date table (one row per day with no gaps or overlaps). In this case, the data is automatically grouped by [Invoice Date Key] because this is the column used in an active relationship. To override the default relationship, it's possible to pass a filter an additional rule. This might take the form of using USERELATIONSHIP or another explicit filter that you would like to apply over the top of the DAX expression for extra flexibility.

The CALCULATE function is not required, nor has there been any use of functions to clear implicit filters. The filter context is being overridden automatically by the TOTALYTD function, allowing the SUM function to see all data it needs to produce a cumulative value.

When this calculated measure is used in a visual that uses other fields from the 'Dimension Date' table, not only does the measure show the appropriately accumulated value for each point in time, it resets back to zero for each year. There is no code in the calculation to instruct the function on how to order the data. This is taken care of automatically because this is a time-intelligence function.

Note   Always note which table any date field you use for the axis on your visual comes from. Using date columns from different tables has an impact on the implicit filters passed to your calculated measures and therefore can help you get the right (or wrong) result.

If you want to use this function but have data to accumulate from a month other than January, you can add an optional parameter. This makes the function useful for fiscal periods that don't start on January 1 each year. If your organization has a fiscal year that ends on the last day of each June, add the value "6/30" as a third parameter. This tells the function to use the 30th day of the 6th month to reset the cumulative measure.

```
My Year to Date (Date Table) =
    TOTALYTD (
        SUM('Fact Sale'[Quantity]),
        'Dimension Date'[Date],
      "6/30"
        )
```

Some functions that are similar TOTALYTD are TOTALMTD and TOTALQTD. These allow you to add calculations that help provide Month-to-Date and Quarter-to-Date totals to your data model.

Period Comparisons

You can use other time-intelligence functions to help make period comparison calculations easier. On a report, it is often helpful to show the value for the same metric for a month immediately prior, or perhaps to show the same calendar month from the previous year next to the value for the current month. These can be extended show values based on the difference—for example, this month is 2 percent up from the previous month or 5 percent down from the same time last year.

Like with the TOTALYTD example, it's possible to build a period comparison calculation that uses the CALCULATE function along with explicit filtering rules, but using a time-intelligence function can make this task easier.

The calculated measure in Listing 7-4 generates a value that shows the value of an expression for the previous month.

Listing 7-4. Previous Month Calculated Measure Using PREVIOUSMONTH

```
QTY for Previous Month =
    CALCULATE(
        [Sum of Quantity],
        PREVIOUSMONTH('Dimension Date'[Date])
        )
```

The calculated measure passes the PREVIOUSMONTH function as a filter to the CALCULATE function to control the data used by the [Sum of Quantity] measure. The PREVIOUSMONTH function returns a table of dates that represents a list of values for each day of the previous calendar month. The function handles any quirks that occur due to consecutive months having a different number of days.

You can now add a simple calculated measure (Listing 7-5) to show a percentage difference between the current and previous month.

Listing 7-5. Percentage Difference Calculated Measure Incorporating the Code in Listing 7-4

```
QTY % Diff to Prev Month =
    DIVIDE(
        [Sum of Quantity] - [QTY for Previous Month],
        [QTY for Previous Month]
        )
```

Figure 7-4 shows these measures added to a table visual in Power BI using the Calendar month field from the 'Date Dimension' table.

Calendar Month	Sum of Quantity	QTY for Previous Month	QTY % Diff to Prev Month
January 2013	193,271		
February 2013	142,120	193,271	-26.47%
March 2013	207,486	142,120	45.99%
April 2013	212,995	207,486	2.66%
May 2013	230,725	212,995	8.32%
June 2013	213,468	230,725	-7.48%
July 2013	232,599	213,468	8.96%
August 2013	192,199	232,599	-17.37%
September 2013	190,567	192,199	-0.85%

Figure 7-4. *Output of code in Listings 7-4 and 7-5*

The [Sum of Quantity] column shows a value relevant to the calendar month for that row. The [QTY for Previous Month] column is a calculation that uses the PREVIOUSMONTH function, whereas the [QTY % Diff to Prev Month] column shows the variation using a percentage.

The [QTY % Diff to Previous Month] calculation could incorporate the logic for both steps in one using variables as is shown in Listing 7-6.

Listing 7-6. Condensed Version of Listings 7-4 and 7-5 Using Variables

```
VAR PrevMonth =
    CALCULATE(
        [Sum of Quantity],
        PREVIOUSMONTH('Dimension Date'[Date])
        )
RETURN
    DIVIDE(
        [Sum of Quantity] - PrevMonth,
        PrevMonth
        )
```

This version stores the result of the CALCULATE function that is using PREVIOUSMONTH as a filter in the PrevMonth variable, which is then used multiple times in the final RETURN statement. Using the output of a variable multiple times can help improve performance.

These are other-time intelligence functions that can be used as filters with CALCULATE:

- PREVIOUSDAY/NEXTDAY

- PREVIOUSQUARTER/NEXTQUARTER

- PREVIOUSYEAR/NEXTYEAR

- SAMEPERIODLASTYEAR

Each of these functions return a table of dates that can be used as a filter parameter for the CALCULATE function. Substituting PREVIOUSMONTH with SAMEPERIODLASTYEAR produces a value from the expression that is filtered to the same calendar month for the previous year.

The DATEADD and PARALLELPERIOD functions also return a table of dates that can be used as a filter in a CALCULATE function. These provide more flexibility than functions such as PREVIOUSMONTH in that they can be configured to jump multiple steps forward or backward to help achieve other reporting requirements.

The calculations in Listings 7-7 and 7-8 use the DATEADD and PARALLELPERIOD functions to produce a value using data going back three months. Both functions have the same parameter signature.

Listing 7-7. Calculated Measure Using PARALLELPERIOD to Look Back Three Months

```
QTY (Month -3) =
    CALCULATE(
        SUM('Fact Sale'[Quantity]),
        PARALLELPERIOD(
            'Dimension Date'[Date],
            -3
            ,MONTH)
        )
```

Listing 7-8. Calculated Measure Using DATEADD to Look Back Three Months

```
QTY (Month -3) =
    CALCULATE(
        SUM('Fact Sale'[Quantity]),
        DATEADD(
            'Dimension Date'[Date],
            -3
            ,MONTH)
    )
```

The Rolling Average

Another common requirement is to show calculations that use data spanning a period relative to the current value, such as a rolling sum/average of the previous *n* number of days, weeks, or months. These measures help provide a smoother view of trends and can eliminate some of the fluctuation noise created by more granular time periods.

For these types of measures to work, the calculations need to override the default filter context with a new filter that is eventually used by the core DAX expression. The objective is to create a table of days that covers the period intended for the calculation.

There are multiple ways to approach this in DAX. Here are a couple of suggestions for creating a calculated measure that shows the rolling average of a daily sum over 'Fact Sale'[Quantity].

The first version is Listing 7-9, which looks back seven days.

Listing 7-9. Rolling Average Calculated Measure Looking Back Seven Days

```
Avg Qty Last 7 Days (Date Table) =
VAR DateFilter =
    DATESINPERIOD(
        'Dimension Date'[Date],
        MAX('Dimension Date'[Date]),
        -7,
        DAY
        )
```

```
VAR RollingSUM =
    CALCULATE(
        [Sum of Quantity],
        DateFilter
        )
RETURN
    DIVIDE( RollingSUM, COUNTROWS( DateFilter) )
```

The DateFilter variable uses the DATESINPERIOD function to create a table that contains the previous seven days including the current day. The MAX function helps to select a single date from what might be many days depending on the context of the calculation when it is executed.

If the calculation is being used on a visual that uses 'Dimension Date'[Date] on an axis, row, or column header, the MAX function only ever needs to pick from a single date. However, if the calculation is being used on a visual that uses the 'Dimension Date'[Calendar Month] on an axis, row, or column header, there will be multiple days for each month. The MAX function simply selects one of the dates, which in this case, is the last day of each calendar month. You could use other functions, such as MIN, LASTDATE, and FIRSTDATE, here in place of MAX.

The table returned by the DATESINPERIOD used in this calculation should never carry more than seven rows.

The next step performs a sum over data in the 'Fact Sale' table that is filtered using the dates stored in the DateFilter variable. Finally, the RETURN statement uses the DIVIDE function to output the final value. The COUNTROWS function uses the number of rows in the DateFilter variable to ensure the six days at the beginning return appropriate averages, rather than simply dividing by seven. However, the COUNTROWS function does not account for days that have no values. The preceding version simply sums all the data it sees in a seven-day period and divides that by seven. In this dataset, there are no invoices generated on Sundays, so a more desirable approach could be to divide the sum of quantity for the period by the number of days with data.

The final RETURN statement for the previous example can be enhanced with an additional step to filter days with no data (Listing 7-10).

Listing 7-10. Updated RETURN Statement Using FILTER for Code in Listing 7-9

```
RETURN
    DIVIDE(
        RollingSUM,
        COUNTROWS(
            FILTER(DateFilter,
                [Sum of Quantity]>0)
            )
        )
```

Or you can use the AVERAGEX function to apply the same effect (Listing 7-11).

Listing 7-11. Alternative Version of Listing 7-9 Using AVERAGEX

```
Avg Qty Last 7 Days (Date Table) =
VAR DateFilter =
    DATESINPERIOD(
        'Dimension Date'[Date],
        MAX('Dimension Date'[Date]),
        -7,
        DAY
        )
RETURN
    AVERAGEX(
        DateFilter,
        [Sum of Quantity]
        )
```

In this case, the AVERAGEX function is an iterator, so the DateFilter variable is passed as the first argument. Remember this variable only contains a single column table of dates with just seven rows. The AVERAGEX function considers that some days have no data and only divides the sum by a divisor appropriate to the underlying data.

Another variation of this uses the DATESBETWEEN function to generate a list of dates to be used to override the calculation (Listing 7-12).

Listing 7-12. Alternative Version of Listing 7-11 Using DATESBETWEEN

```
Avg Qty Last 7 Days (Date Table) =
VAR DateFilter =
    DATESBETWEEN(
        'Dimension Date'[Date],
        LASTDATE('Dimension Date'[Date])-6,
        LASTDATE('Dimension Date'[Date])
        )
RETURN
    AVERAGEX(
        DateFilter,
        [Sum of Quantity]
        )
```

The DATESBETWEEN function differs from DATESINPERIOD in that it uses two fixed dates to determine the date range rather than using a single date with a relative offset. The DATESINPERIOD function has the advantage of being able to specify offset periods such as year, quarter, and month as well as day.

Rolling Your Own Table

Having a dedicated, centralized date table in your data model is not only useful for filtering and grouping data across multiple tables in unison, but it can also help you provide a stable set of dates for time intelligence functions to make use of.

There are multiple ways to add a date table to a data model. You can import date data from an existing data source that already carries one row per day for a date range that covers all the data in your model, or you can generate your own date table using the query editor or DAX.

This section shows you how to build a data table from scratch using DAX functions without needing to rely on an external data source. You do not need to have a version of DAX that supports calculated tables.

The starting point is a DAX function that can generate a table with the appropriate number of rows. The two functions that help with this are CALENDAR and CALENDARAUTO. Both generate a contiguous series of dates in a single column table.

CALENDAR

You can use the CALENDAR function when you know the start and end dates for the date range required for your model.

The syntax for this function is

```
CALENDAR ( <startDate>, <endDate> )
```

Some suggested variations for using this function are as follows:

```
Dates = CALENDAR("2016-01-01", "2019-12-31" )
```

```
Dates = CALENDAR(DATE(2016,1,1), TODAY())
```

```
Dates = CALENDAR(
          FIRSTDATE('Fact Sale'[Invoice Date Key]),
          LASTDATE('Fact Sale'[Invoice Date Key])
          )
```

```
Dates = CALENDAR(
          STARTOFYEAR( 'Fact Sale'[Invoice Date Key] ),
          ENDOFYEAR( 'Fact Sale'[Invoice Date Key] )
          )
```

The output for each of these statements is a table with a single column called [Date] that you can use as the basis for adding columns using additional DAX expressions.

CALENDARAUTO

A handy alternative to the CALENDAR function is CALENDARAUTO. The CALENDARAUTO function does not require parameters. To generate its own start and end dates, it looks at existing tables in the data model for columns that use Date or DateTime datatypes. If it finds them, it finds the newest and oldest values before rounding out to the start and end of each calendar year.

Any calculation using CALENDARAUTO is re-executed each time data is refreshed, so if new data arrives in your data model that falls outside the existing date boundary, rows in a table using this function are added automatically.

Expanding the Date Table

The first columns I normally add to a date table provide the ability to slice and dice by calendar month. You can add these using the ADDCOLUMNS function, which adds columns to a base table, each as a pair of parameters that specifies the name of the new column along with the DAX expression to be used to generate the value for each row of that column (Listing 7-13).

Listing 7-13. Calculated Table Using ADDCOLUMNS to Build on CALENDAR

```
Dates =
VAR BaseTable = CALENDAR("2016-01-01",TODAY())
RETURN
    ADDCOLUMNS(
        BaseTable,
        "MonthID", FORMAT([Date],"YYYYMM"),
        "Calendar Month", FORMAT([Date],"MMMM YY")
        )
```

Let's call this calculated table Dates. The CALENDAR function is used to generate a single column table that has a contiguous sequence of dates beginning on January 1, 2016, that run to the current day and is assigned to the BaseTable variable. This table grows by one row each day.

Two columns named MonthID and Calendar Month are added using the ADDCOLUMNS function. The MonthID column is added so the text-based Calendar Month column can be sorted chronologically. Both columns use the FORMAT function to convert the intrinsic date value stored in the [Date] column to a value useful for each column.

The FORMAT function converts numbers or dates into text values. Characters, symbols, and other text can be added as part of the conversion. The FORMAT function can be used with format string rules to help enforce the number of decimals to be displayed.

FORMAT can also convert date values to text using predefined or custom formatting rules. In this case, both columns added to the Date table opt for custom strings. Uppercase M characters represent the underlying month. Two M characters return a two-character number padded with a leading zero. A date that falls in January returns "01", whereas a date that falls in December returns "12". The uppercase Y represents a year component. Four Y characters signify that the full year including the century should be generated as text. Two Y characters return the two least significant values from the year, so 1918 and 2018 both return "18".

The FORMAT function used for the MonthID column uses a format string of "YYYYMM". When this format string is applied to a date such as the January 5, 2019, the output is "201901". The output for July 4, 2019, is "201907". The text data in this column is now sorted chronologically for date. This means that "202008" always appears later than "199910", so you can use this column to control the sort order of the Calendar Month column.

The format string used by the FORMAT function for Calendar Month is "MMMM YY". Four uppercase M characters signify that the month name should be used in full. The space character used in the format string also flows through to the final output. Hyphens and other dashes or separators can also be used here. Finally, the two uppercase Y characters are used to ensure that data across years is not grouped together. The output of this function for January 5, 2019, would be "January 19", whereas July 4, 2019, would be "July 19".

Note   When using text to represent a calendar month, it is good practice to include extra text to also uniquely identify the year. Otherwise you risk having January 2019 data merged with January 2020 in the same calculation.

Once the calculated table has been added to the model, the Calendar Month column can be set to use the MonthID as its Sort By column. You can hide the MonthID column and create relationships between this table and other tables in the model.

An alternative version of the MonthID column involves using a DateTime value rather than an integer datatype (Listing 7-14). When DateTime fields are used on the axis of many of the visuals in Power BI, labels are dynamically scaled so no scrolling is required. This also allows the column to use date-based functions in additional calculations. The code shown in Listing 7-14 ensures every value for the "Calendar Month" column still only has one unique value rows for the MonthID column. The [Date] value for the "Month" column is converted to be the first day for each month using the DATE, YEAR, and MONTH functions.

Listing 7-14. Using a [Date] Value in Place of Text for the Code in Listing 7-13

```
RETURN
    ADDCOLUMNS(
        BaseTable,
        "Month", DATE(YEAR([Date]),MONTH([Date]),1),
        "Calendar Month", FORMAT([Date],"MMMM YY")
        )
```

Adding the Year

To add a column for the calendar year, use the parameters with the ADDCOLUMNS function (Listing 7-15).

Listing 7-15. Adding a Year Column to the Calculated Table

```
RETURN
    ADDCOLUMNS(
        BaseTable,
        "Month", DATE(YEAR([Date]),MONTH([Date]),1),
        "Calendar Month", FORMAT([Date],"MMMM YY"),
        "Year", FORMAT( [Date], "YYYY" )
        )
```

This adds a text column called "Year" that is already sorted chronologically so it has no need for another column to be generated to fix sorting issues. I have come across an interesting requirement to force the sorting of years to be in reverse. This was to control the order in which values in this column appear when they are used as a column header on a matrix. The requirement was to show later years on the left with older values on the right. To solve this issue, you can add the following helper column to the model:

```
"YearSortID-", 0 - YEAR( [Date] )
```

This produces a negative version of the year, meaning that when a Year column is sorted by this column, the order is flipped so 2019 appears after 2020.

Fiscal Year

To add a column to carry a value for the fiscal year, one approach is to add the expression in Listing 7-16 to the ADDCOLUMNS statement.

Listing 7-16. Name and Expression to Add a Column Called "Fiscal Year" to the ADDCOLUMNS Function

```
"Fiscal Year",
    VAR FY_Month_Starts = 6
                RETURN YEAR([Date]) - IF(
                                    MONTH([Date]) < FY_Month_
                                    Starts,
                                    -- Then Add a 1 to the
                                    year --
                                    1,
                                    -- Else leave as is  --
                                    0  )
```

This statement creates a column named "Fiscal Year". The expression declares a variable called FY_Month_Starts, which is assigned an integer between 2 and 12. A fiscal year starting in June would use 6. The RETURN statement begins with a value for the current calendar year and then subtracts a year depending on the logic in the IF statement.

This calculation outputs the same value for both Fiscal and Calendar year for dates after June 1. Dates between January 1 and the end of May would automatically retain the value for the previous year.

To create a column that shows a month number based on a fiscal calendar, you can add the calculation in Listing 7-17.

Listing 7-17. Name and Expression to Add a Column Called "Fiscal Month No" to the ADDCOLUMNS Function

```
"Fiscal Month No",
    VAR FY_Month_Starts = 6
    RETURN MOD(
        MONTH([Date]) - FY_Month_Starts,
        12
        )+1
```

189

The value assigned to the FY_Month_Starts variable represents the calendar month number to use as a starting point. This calculation produces a value of 1 for June, 2 for July, 3 for August, through to 12 for May.

Days from Today

A handy column to have in any date table is one that carries a number, negative or positive, that shows how many days each row is in relation to the current day. This column can then be used to provide date-based filters that are relative and dynamic. A suggested calculation for this is

```
"Days from Today",INT( [Date] - TODAY())
```

This column carries a value of 0 for the row that has a value in the [Date] column that matches the current day. Values in rows that have future values have positive numbers, whereas date values in the past have negative numbers.

To use this column to enforce a dynamic view of the last 14 days of data, add this field to a Report, Page, or Visual filter and assign an advanced rule saying it should only show values between –14 and 0. Once set, the filter setting can be left alone, and the report always shows the most recent 14 days.

Weekly Buckets

Weeks can be tricky columns to work with and generally they don't play nicely with other date-based groups such as months and years. It's common for weeks to be split across the boundaries of month or year buckets so often that they need to be used in isolation from other types of date groupings.

Weeks are still a useful way to organized dates in a model, however. An effective way to do this is to group batches of seven days using a value that is either the first or last date of the week. The WEEKDAY function is useful to help achieve this because it assigns each [Date] a value between 1 and 7 depending on which day of the week it is. The calculation for a week starting column might look like this:

```
"Week Starting Sunday", [Date] - ( WEEKDAY( [Date] ) -1 )
```

By default, the WEEKDAY function assumes a week begins on a Sunday, so it assigns the value of 1 to any date that happens to be a Sunday, 2 to Monday, 3 to Tuesday, and so on until Saturday which is assigned 7. The calculation then offsets each value by 1 so it becomes zero based. The new value is then subtracted from [Dates], which now aligns all dates to the closest Sunday they follow.

A calculation that uses Monday instead of Sunday as the basis for a week starting column is the following:

```
"Week Starting Monday", [Date] - WEEKDAY( [Date], 3 )
```

The WEEKDAY function now has a second parameter that instructs the function to return a value between 0 and 6 with Monday being 0, Tuesday being 1, Wednesday being 2, and so on through to Sunday which is 6. This output is already zero based so you no longer need to subtract 1 from the result.

A slightly trickier scenario is one in which the week needs to start or finish based on a day other than a Sunday or Monday. In this case you can add the code in Listing 7-18.

Listing 7-18. Name and Expression to Add a Column Called "Week Starting Wednesday" to the ADDCOLUMNS Function

```
"Week Starting Wednesday",
        VAR myWeekDay = WEEKDAY( [Date], 3 )
                    VAR offset = 2 - 0=Mon, 1=Tues,
                    2=Wed, 3=Thur, 4=Fri 5=Sat, 6=Sun
                    RETURN [Date] - MOD( myWeekDay -
                    offset, 7 )
```

You can configure this expression so any day can be the starting day for the week. The value assigned to the offset variable controls which day to use. In this case, it has been set to 2, which means Wednesday is the starting day for each week. For Saturday, you would assign a value of 5 to the offset variable.

Once these DateTime columns are in place, you can add additional columns that count the number of weeks to or from a milestone. The current day is again a good milestone to work from, so a column that shows "Weeks from today" might look like this:

```
"Weeks from today",
    DIVIDE(
        INT(
            ([Date] - WEEKDAY([Date],3)) - TODAY()
            ),
        7)
```

This calculation includes the same logic used in the "Week starting Monday" column, but it extends to find the number of actual days between the starting date for the week and the current day (TODAY). The calculation then divides by seven to provide a zero-based index that shows negative numbers for dates in the past and positive numbers for future dates.

Is Working Day

Two columns that can help represent working days can be a simple column that shows just 1 or 0 values that reflect if the specific row happens to be a working day or not and a column that shows the number of working days from a point (such as the current day).

The first example (Listing 7-19) assumes that Monday through Friday should be marked as 1 to represent a working day, while Saturday and Sunday should be marked with a 0 to represent a weekend date. You can reverse this if you intend to show weekends as 1 and weekdays as 0.

Listing 7-19. Name and Expression to Add a Column Called "Is Work Day" to the ADDCOLUMNS Function

```
"Is Work Day",
    IF(  -- Zero based index with Monday being 0 is less than 5
            WEEKDAY( [Date], 3 ) < 5,
            -- THEN --
            1,
            -- ELSE --
            0
            )
```

The WEEKDAY function returns a value between 0 and 6 to each [Date] when used with the optional second parameter of 3. Mondays = 0, Tuesdays = 1, and on through to Sunday, which is 6. The IF statement simply tests the output of the WEEKDAY function and assigns the appropriate value.

A second useful column (Listing 7-20) is one that shows the number of working days since a specific point in time. This example uses the current day as the point to count from and dynamically updates each day.

Listing 7-20. Calculated Column Showing the Number of Working Days from the Current Date.

```
Working Days From Today =
VAR ColDate = 'Dates'[Date]
RETURN
    IF( ColDate < TODAY(),
      -- THEN --
        0-CALCULATE(
            SUM('Dates'[Is Working Day]),
            ALL(Dates),
            'Dates'[Date] >= ColDate &&  'Dates'[Date] < TODAY()
            )
      -- ELSE --
            ,
        CALCULATE(
            SUM('Dates'[Is Working Day]),
            ALL(Dates),
            'Dates'[Date] <= ColDate &&  'Dates'[Date] > TODAY()
            )
            )+0
```

This calculation can only be added to an existing physical Date table. It cannot be added as part of the ADDCOLUMN function used in a create table statement since the ALL function relies on a table that must already exist.

The calculation also relies on having a column called [Is Working Day] that already exists on the table. I included a suggested version of this earlier in the "Is Working Day" section. The calculation stores the value from the 'Date' column in a variable called ColDate. This can only ever be one value at a time. The IF statement then decides which of the two CALCULATE statements should be used. The first is designed to handle all the rows older than the current date, while the second CALCULATE applies a slightly different logic to calculate the value for rows newer than the current date. Only one of the two CALCULATE functions runs per row.

You could compress this into a single CALCULATE statement, but I have presented it like this to show you the general approach. Both CALCULATE statements use the ALL function to unlock rows other than the specific row being calculated. The SUM function can then use data from a wider range of rows to produce the result.

193

These calculations do not take public holidays into account since holidays can vary wildly from region to region and even year to year. A suggested approach to handling public holidays is to keep a separate table that carries a row for every instance. This can be maintained manually or be sourced from various local datasets. You can then use data from such a table in calculations to help determine if a row in your actual date table falls on a public holiday.

Weekday Name

You can use the FORMAT function to show a column that simply carries the name of the day of the week. The DAX code for adding this using ADDCOLUMNS follows:

```
"Weekday name", FORMAT([Date],"DDDD")
```

There are four uppercase D characters used here in the format string to specify that the weekday should be shown in full, such as "Monday" or "Tuesday". Using three uppercase D characters returns an abbreviated version such as "Mon" or "Tue".

Rolling Your Own—Summary

It is easy enough to generate a Date table for your model using only DAX. You can use a mixture of these suggestions to form the basis of a date table suitable for your data model. The examples are designed to give you an idea of how you might approach creating columns that you can tweak to suit your own organization's requirements.

Optimizing Dates

Raw data from source systems often carries data to represent a value that points to a specific date and time. This is quite common when you're dealing with sensor, Internet of Things (IOT), or transactional datasets. These may carry a field that shows the year, month, day, hour, minute, and second. A good practice to follow when using this type of data is to split the value into two or more columns.

One column should be Date only, so it should be truncated to only show the year, month, and day. The second column (if you need it at all) can carry a version of the time. The time can be truncated to just a number between 0 and 23 to represent the hour of the day, or it can be a number between 0 and 1,440, which represents the specific minute of the day the event took place.

The reason for doing this is to help optimize the data model for speed. When the DAX engine imports data, it splits tables of data to individual columns. The data in each column is then compressed to unique values. If you have a field in your raw data that carries a DateTime field down to the second (or microsecond), the column is highly unique and therefore does not compress much.

If your raw data is an output from an IOT sensor that creates five readings every second, and you import this data "as is" to your model, it has the potential to create 157 million unique values per year that cannot be compressed (5 readings per second * 60 seconds * 60 minutes * 24 hours * 365 days = 157,680,000).

If the same sensor data is split into two columns, one column for the Date component and the other column for showing a value for the minute of the day, the Date column will have just 365 unique values whereas the column representing the minute will have just 1,440. This approach means the raw data can be highly compressed when imported and it is still possible to perform the same calculations using this much smaller and faster data model.

CHAPTER 8

Debugging and Optimizing

Debugging in DAX

A good question in any language is, "How can I debug the code?" Some languages have sophisticated tools that provide the programmer with the ability to walk step by step through execution of the code as the application runs. These debug tools often allow you to pause at various points to inspect all nature of useful properties and values so you get a clearer idea on the actual flow of logic as it pans out. This is not always the path the you expect, but at least being able to debug can help you identify sections of code you need to improve.

Unfortunately, DAX doesn't have such a sophisticated tool available to debug with, but fortunately it also doesn't have the same degree of logical pathways that you might expect in even a simple application. Calculations in DAX are often single-line statements that are as simple as passing just one parameter to a function.

Step debugging the following calculated measure probably won't provide you with much useful information, but the output is quick to generate and enough to help you understand the reason behind the result.

```
Count of Table = COUNTROWS( 'Table' )
```

Given that calculated measures can be executed many times inside a report page, it can be difficult to isolate the execution of a specific instance to understand what filter context may or may not be at play for that case.

That said, although you can't start debugging or step and run through your code, you can apply several types of manual techniques to your code while you're building calculations. This is what I like to think of as old-school debugging.

© Philip Seamark 2018
P. Seamark, *Beginning DAX with Power BI*, https://doi.org/10.1007/978-1-4842-3477-8_8

The general principal is to break calculations out into separate, smaller versions and use the output of each smaller calculation to help understand the values being generated. By reducing the logic down to smaller baby steps, you can inspect each value to see which ones match your expectations and which may be different. Tracing the baby steps through to the first point where they generate unexpected results should be the fasted way to show which aspect of code needs your attention.

If you have a calculation generating unexpected results, make a copy and change the core expression to something simpler, such as a COUNTROWS function. Doing this can often help you figure out how many rows are being considered for the calculation.

In the case of a tricky calculated column that consists of several nested functions, create some temporary columns that are subsets of each function.

To walk through an example, take a look at the calculated column in Listing 8-1; this calculation shows a percentage ratio for each product sold per day.

Listing 8-1. Calculated Column Creating a Ratio of Sales Against a Grand Total

```
Product Ratio of Daily Sales =
DIVIDE (
    CALCULATE (
            SUM('Fact Sale'[Total Amount]),
            ALL('Fact Sale'),
            'Fact Sale'[Invoice Date Key]=EARLIER('Fact Sale'[Invoice Date Key]),
            'Fact Sale'[Stock Item Key]=EARLIER('Fact Sale'[Stock Item Key])
    )
    ,
    CALCULATE (
            SUM('Fact Sale'[Total Amount]),
            ALL('Fact Sale'),
            'Fact Sale'[Invoice Date Key]=EARLIER('Fact Sale'[Invoice Date Key])
    )
)
```

The first CALCULATE in this statement attempts to determine the total value for the specific product for each day. Because there may be other transactions on the same day for the same product, the calculation needs to check to see if there are other rows in the same table with the same [Stock Item Key] and [Invoice Date Key]. If it can't find any, and there is only one transaction per day for the product, this value should exactly match the value in the [Total Amount] column. If other rows are found for the same [Stock Item Key] and [Invoice Date Key], the SUM function should generate the correct value for all transactions combined and then assign that value to each instance of that product.

The only values that can be seen are the numbers returned as the final output of the entire calculated column. Intermediate values such as those generated by the first CALCULATE function are lost.

The second CALCULATE attempts to derive a total value for sales for each day. This daily total is used with the first value to derive a percentage for each product/day combination. Once again, the way the calculation has been written, you do not see these intermediate values. How can we you sure the values used in the DIVIDE calculation are the values you think they should be?

The easiest way to break this calculation down is to create a separate observable entity for each of the CALCULATE functions. This can take the form of several new calculated columns, or you can use variables and control which variable is returned via the RETURN statement.

Debugging Using Columns

Using the calculated column approach, you can add the three calculated columns in Listings 8-2, 8-3, and 8-4.

Listing 8-2. Calculated Column Showing the Sub Category Total

```
Product Sub Total =
    CALCULATE (
            SUM('Fact Sale'[Total Amount]),
            ALL('Fact Sale'),
            'Fact Sale'[Invoice Date Key]=EARLIER('Fact Sale'[Invoice Date Key]),
            'Fact Sale'[Stock Item Key]=EARLIER('Fact Sale'[Stock Item Key])
        )
```

Listing 8-3. Calculated Column Showing the Overall Total

```
Daily Sub Total =
        CALCULATE (
            SUM('Fact Sale'[Total Amount]),
            ALL('Fact Sale'),
            'Fact Sale'[Invoice Date Key]=EARLIER('Fact Sale'[Invoice Date Key])
            )
```

Listing 8-4. Calculated Column Incorporating Listings 8-2 and 8-3

```
Divide Test =
    DIVIDE(
        'Fact Sale'[Product Sub Total],
        'Fact Sale'[Daily Sub Total]
        )
```

In these listings, the two CALCULATE functions from Listing 8-1 have been separated into their own calculated columns while a third calculated column uses the DIVIDE function that incorporates the other calculated columns. You can visually inspect this in the Data View as a form of debugging. You can use these three new calculated columns in addition to the original version of the calculated column.

Figure 8-1 shows a sample of the table with the three new "debug" columns. The table has been filtered to only show data for January 1, 2016, and it is ordered by the [Stock Item Key] column.

Invoice Date Key	Stock Item Key	Total Amount	Product Ratio of Daily Sales	Product Sub Total	Daily Sub Total	Divide Test
1/01/2016 12:00:00 AM	1	750	5.40%	750	13888	5.40%
1/01/2016 12:00:00 AM	9	49.2	0.44%	61.5	13888	0.44%
1/01/2016 12:00:00 AM	9	12.3	0.44%	61.5	13888	0.44%
1/01/2016 12:00:00 AM	14	19.44	0.14%	19.44	13888	0.14%
1/01/2016 12:00:00 AM	16	240	1.73%	240	13888	1.73%
1/01/2016 12:00:00 AM	17	432	3.11%	432	13888	3.11%
1/01/2016 12:00:00 AM	20	118.08	1.59%	221.4	13888	1.59%
1/01/2016 12:00:00 AM	20	103.32	1.59%	221.4	13888	1.59%
1/01/2016 12:00:00 AM	21	106.56	0.77%	106.56	13888	0.77%

Figure 8-1. *Output of the code in Listings 8-2, 8-3, and 8-4*

The [Product Sub Total] value of 750 for the first row seems to match the [Total Amount] column. There is only one row for [Stock Item Key] = 1, so this looks good. The next two rows have been highlighted in a box. This shows that there were two transactions on this day for [Stock Item Key] = 9. The individual totals of 49.2 and 12.3 combine for a total of 61.5. This is the value repeated for both rows in the [Product Sub Total] column, which is the desired result. The number still to be tested is the value of 13,888 that is repeated in every row of the [Daily Sub Total] column.

One approach to testing this is to first create a calculated measure that uses SUM of the [Total Amount] column and then use this calculation in a visual where a filter is set to the same day as the test data. If the source system is a SQL database, another option is to write a query to help reconcile this value. Ideally T-SQL queries should be kept simple, otherwise the test can be compromised by errors in the T-SQL statement, so avoid joins and stick to simple SUM and COUNT functions with just the bare minimum of code in any WHERE clause.

Debugging Using Variables

An alternative to the preceding approach is to make use of variables. The object is to break the calculation into smaller chunks of code wherever practical. The same code split up into variables might look like Listing 8-5.

Listing 8-5. Calculated Measure Using RETURN to Control the Final Output Variable

```
Product Ratio of Daily Sales =

VAR InvoiceDateKeyCol = 'Fact Sale'[Invoice Date Key]
VAR StockItemKeyCol = 'Fact Sale'[Stock Item Key]

VAR ProductSubTotal =
    CALCULATE (
            SUM('Fact Sale'[Total Amount]),
            ALL('Fact Sale'),
            'Fact Sale'[Invoice Date Key]=InvoiceDateKeyCol,
            'Fact Sale'[Stock Item Key]=StockItemKeyCol
        )
```

```
VAR DailySubTotal =
    CALCULATE (
        SUM('Fact Sale'[Total Amount]),
        ALL('Fact Sale'),
        'Fact Sale'[Invoice Date Key]=InvoiceDateKeyCol
        )

VAR ReturnValue =
    DIVIDE (
        ProductSubTotal,
        DailySubTotal
        )

RETURN  ProductSubTotal
```

In Figure 8-5, I have modified the original calculated measure to use variables to store the various sections of code. Apart from the fact that this is now more readable, the final RETURN statement is not obliged to return the last variable to be declared.

The RETURN statement here returns the ProductSubTotal and not the ReturnValue variable, meaning once this calculated column is executed, numbers visible in this column in the Data View represent the output of the first CALCULATE function and not the output of the final DIVIDE statement.

When these values have been checked and confirmed to be as expected, you can update the final RETURN statement so it returns the DailySubTotal variable. When this has been confirmed as being correct, you can set the RETURN statement to use the ReturnValue column.

The downside of this approach is that only one variable can be returned at any one time. You can take a hybrid approach if one of the calculated columns used earlier produces a value that might be useful in several other calculations, however. You can use the [Daily Sub Total] column to create similar "percentage of"–type calculations that involve other columns in the table.

Debugging Calculated Measures

When it comes to calculated measures, the principle is the same. The calculated measure in Listing 8-6 meets the same product ratio of daily sales requirement in a single calculation.

Listing 8-6. Calculated Measure to Be Debugged

```
Product Ratio of Daily Sales as Measure =
    DIVIDE (
        CALCULATE (
                SUM('Fact Sale'[Total Amount]),
                FILTER(
                ALLSELECTED('Fact Sale'),
                'Fact Sale'[Invoice Date Key]=MAX('Fact Sale'[Invoice Date
                Key]) &&
                    'Fact Sale'[Stock Item Key]=MAX('Fact Sale'[Stock Item Key])
                    )
            )
            ,
        CALCULATE (
            SUM('Fact Sale'[Total Amount]),
            FILTER(
                ALLSELECTED('Fact Sale'),
                'Fact Sale'[Invoice Date Key]=MAX('Fact Sale'[Invoice
                Date Key])
                )
            )
        )
```

Listings 8-7, 8-8, and 8-9 show how to debug this DAX by breaking it into three separate calculated measures.

Listing 8-7. Calculated Measure Showing Total for Sub Category

```
Product Sub Total as Measure =
        CALCULATE (
                SUM('Fact Sale'[Total Amount]),
                FILTER(
                ALLSELECTED('Fact Sale'),
                    'Fact Sale'[Invoice Date Key]=MAX('Fact Sale'[Invoice
                    Date Key]) &&
                    'Fact Sale'[Stock Item Key]=MAX('Fact Sale'[Stock Item Key])
                    )
                )
```

Listing 8-8. Calculated Measure Showing Overall Total

```
Product Sub Total as Measure =
        CALCULATE (
                SUM('Fact Sale'[Total Amount]),
                FILTER(
                ALLSELECTED('Fact Sale'),
                    'Fact Sale'[Invoice Date Key]=MAX('Fact Sale'[Invoice
                    Date Key]) &&
                    'Fact Sale'[Stock Item Key]=MAX('Fact Sale'[Stock Item Key])
                    )
                )
```

Listing 8-9. Calculated Column Incorporating Listings 8-7 and 8-8

```
Product Ratio as Measure =
    DIVIDE(
        [Product Sub Total as Measure],
        [Daily Sub Total as Measure]
        )
```

Because these are calculated measures, you need to add them to a visual such as the table visual in Figure 8-2 in order for the values to be inspected. In this figure, the [Invoice Date Key], [Stock Item Key], and [Total Amount] columns have been added to the same visual to provide the calculated measures with some query context.

Invoice Date Key	Stock Item Key	Total Amount	Product Sub Total as Measure	Daily Sub Total as Measure	Product Ratio as Measure
1 January 2016	1	750.00	750.00	13,887.90	5.40%
1 January 2016	9	12.30	61.50	13,887.90	0.44%
1 January 2016	9	49.20	61.50	13,887.90	0.44%
1 January 2016	14	19.44	19.44	13,887.90	0.14%
1 January 2016	16	240.00	240.00	13,887.90	1.73%
1 January 2016	17	432.00	432.00	13,887.90	3.11%
1 January 2016	20	103.32	221.40	13,887.90	1.59%
1 January 2016	20	118.08	221.40	13,887.90	1.59%
1 January 2016	21	106.56	106.56	13,887.90	0.77%

Figure 8-2. *Output of the code in Listings 8-7, 8-8, and 8-9*

The final three columns in Figure 8-2 show the results of each calculated measure. You can use this format to help reconcile against the value for [Total Amount]. The rows in which [Stock Item Key] = 9 add up correctly for the [Product Sub Total as Measure] values.

Keeping the calculation as a single statement but using the RETURN statement to control which variable to return works just as well. The advantage of splitting out to separate calculated measures is that you can then use each measure in other calculations. This introduces a degree of code fragmentation when you are trying to follow code that uses multiple calculated measures, however.

A common reason why calculated measures generate unexpected results is incorrect filtering. As calculated measures grow in complexity, particularly those that mix the use of clearing or applying layers of filtering of the data, this means the core expression doesn't use the data as you expect it to.

Sometimes this becomes obvious when each component is broken out into smaller chucks in the way as shown earlier in this chapter at Listing 8-6. If this still doesn't clarify why a value might be different than what you are expecting, using alternative functions in your core expression may help shed light on the issue.

Taking a copy of your calculated measure and substituting SUM ('Fact Sale'[Total Amount]) with COUNTROWS ('Fact Sale') can sometimes be enough to help identify a filter-related issue. Otherwise substituting with other expressions to show the MIN or MAX of a column, such as MAX ('Fact Sale'[Invoice Date Key]), might be enough to help clarify any issues with filtering.

Some additional functions that can be useful when debugging manually are these:

- HASONEFILTER: Returns true when the specified column has one and only one direct filter value.

- HASONEVALUE: Returns true when there's only one value in the specified column after it is cross filtered.

- ISCROSSFILTERED: Returns true when the specified table or column is cross filtered.

- ISFILTERED: Returns true when there are direct filters on the specified column.

These functions produce true/false outputs and can be useful when you add them to new calculated measures and to the same table visual you are using to debug with. These provide useful information about the level of filtering that the engine is applying. When you are happy your code is running as expected, you can remove these temporary calculated measures.

A final tip is to use a blank report page for debugging and start with just the data you need. If you try to debug measures on a busy report page that has other visuals and filters, it can be tricky to isolate an issue. Once the calculation is working as you expect it to, you can remove or hide the report page you used to debug.

External Tools

Whether you are writing DAX calculations in Power BI, Excel, or SSAS, there are several useful tools that you can use to help produce better calculations. These are three that we will look at in this chapter:

- DAX Studio
- SSMS (SQL Server Management Studio)
- SQL Server Profiler

These tools are all free, and if you are already working with Microsoft SQL Server databases, you probably already have SSMS and SSMS Profiler installed. This section shows you how to get started and highlights key ways each tool can be helpful.

DAX Studio

The first tool to look at is the highly recommended DAX Studio, which describes itself as the ultimate tool for working with DAX queries. You can find documentation on DAX Studio including a link to download from `daxstudio.org`.

Once DAX Studio is downloaded and installed, you can create connections to data models hosted by the following engines:

- PowerPivot in Excel
- Power BI Desktop
- Azure Analysis Services
- Azure Analysis Services Tabular

This means that if you have a copy of Power BI Desktop open with a data model loaded with data, you can open DAX Studio and create a connection to the Power BI Desktop model and start issuing queries.

The first dialog you encounter when opening DAX Studio is one that asks which data model you would like to connect to (Figure 8-3). If you have multiple copies of Power BI Desktop, you can select which one you would like to connect to from a drop-down list.

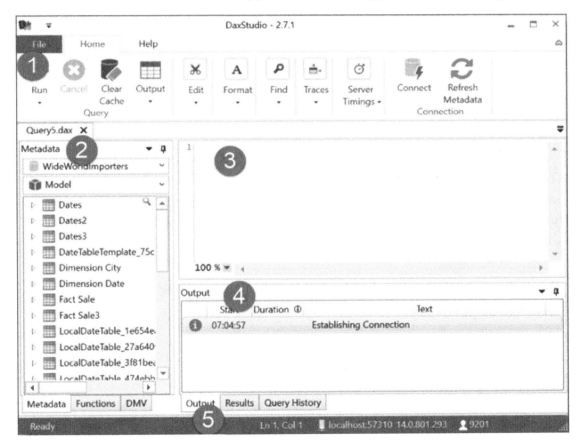

Figure 8-3. *The connection dialog in DAX Studio*

Once connected to a model, the application should appear as it does in Figure 8-4.

Figure 8-4. *The main screen of DAX Studio*

1. The ribbon across the top is loaded with features you can use for editing and optimizing your DAX; you can hide this to provide more space to write DAX.

2. The Metadata panel allows you to explore the structure of your model. You can drag tables and columns to the query editor to save yourself from having to type longer object names. Every DAX function is listed in this panel along with some interesting DMV (dynamic management views) queries.

3. The query editor is where you can write and edit your DAX queries. This features full IntelliSense that suggest functions, tables, columns, and measures as you type.

4. The Output bar consists of several tabs that contain information regarding any query that has been run. The Results tab shows the output of the query. You can enable other tabs to see query plans and internal server timings.

5. The Status bar along the bottom shows information about the current connection.

Once you connect to a data model, you can start writing DAX in the query editor. When you are ready to run your query, click the Run button on the ribbon or press F5. As long as your query has no errors, you can see the output in a grid in the Results tab of the Output panel. There is an option on the ribbon to output the results to a file, which can be a handy way to manually extract data from your data model.

Note   Every DAX query run in DAX Studio must begin with the keyword EVALUATE and must output a table.

Taking the code from a calculated measure and running it in DAX Studio produces an error. The following statements will both fail to run:

```
My Measure = COUNTROWS('Fact Sale')
COUNTROWS('Fact Sale')
```

Even adding the EVALUATE keyword in front of both statements produces an error because the output of the COUNTROWS function is not a table.

```
EVALUATE
    COUNTROWS('Fact Sale')
```

To get a query like this using the COUNTROWS function to work, you need to wrap it in a function that can output to a table such as ROW.

```
EVALUATE
    ROW(
        "My Value",
        COUNTROWS('Fact Sale')
    )
```

This statement produces the following (Figure 8-5) in the Results tab of the Output panel.

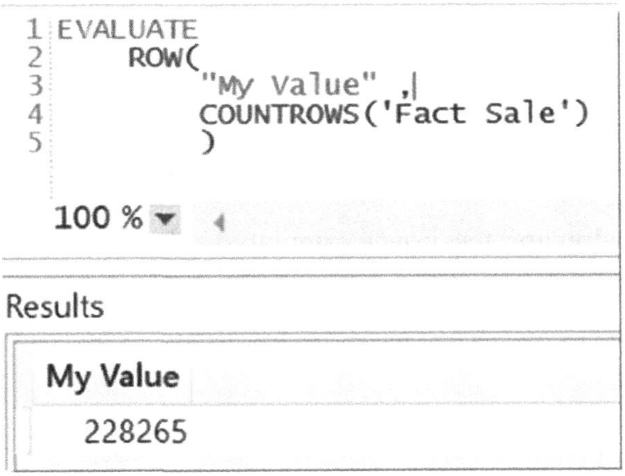

Figure 8-5. *Simple query in DAX Studio including results*

One of the simplest queries to run in DAX Studio is to use EVALUATE with the name of a table in the model.

```
EVALUATE
    'Dimension Date'
```

This is the equivalent of SELECT * FROM [Dimension Date] and shows every row and every column of the 'Dimension Date' table. To reduce the number of rows in the output, you can use filter functions with the table.

```
EVALUATE
    FILTER(
        'Dimension Date',
        'Dimension Date'[Calendar Year]=2016
        )
```

The preceding calculation returns all columns from the 'Dimension Date' table but only rows that belong to [Calendar Year] = 2016.

The following calculation is the equivalent of SELECT TOP 10 * FROM [Dimension Date] ORDER BY [Date] Desc:

```
EVALUATE
    TOPN(
        10,
        'Dimension Date',
        'Dimension Date'[Date],
        DESC
        )
```

If you want to develop and test how a calculated column might look, you can use syntax such as this:

```
EVALUATE
    SELECTCOLUMNS(
        'Fact Sale',
        "My Test Column",
        'Fact Sale'[Unit Price] * 'Fact Sale'[Quantity]
        )
```

The SELECTCOLUMNS function used in this example means the output only contains one column; however, it produces as many rows as there are in the 'Fact Sale' table. You can add additional columns as needed, or, if it is more useful to see every column from your base table including your additional calculation, using the ADDCOLUMNS function might be better (Listing 8-10).

Listing 8-10. DAX Query with an Additional Column Using Nested Variables

```
EVALUATE
VAR DoubleMe = 2
RETURN
    SELECTCOLUMNS(
        'Fact Sale',
        "My Test Column",
        'Fact Sale'[Unit Price] * 'Fact Sale'[Quantity],
        "My Test Column 2",
         VAR X = COUNTROWS('Fact Sale') * DoubleMe
        RETURN X
        )
```

Expanding the logic to include additional columns can be done as shown in Listing 8-10. This also demonstrates how you can use variables both before and inside a query. Figure 8-6 shows the results of Listing 8-10.

My Test Column	My Test Column 2
13	456530
13	456530
13	456530
13	456530
13	456530
13	456530
12	456530

Figure 8-6. *Output of the code in Listing 8-10*

An option to help test calculated measures is to use the DEFINE MEASURE statement to keep DAX logic for the measure separate from the EVALUATE query.

Listing 8-11 creates a new calculated measure that exists only in the scope of this query and is not visible to anyone else using the data model.

Listing 8-11. DAX Query Using DEFINE MEASURE to Create Query-Scope
Calculated Measure

```
DEFINE MEASURE
    'FACT Sale'[My test measure] = COUNTROWS('Fact Sale')

EVALUATE
    SUMMARIZECOLUMNS(
        'Dimension Date'[Calendar Year],
        "Column Name here",
        [My test measure]
        )
    ORDER BY
        'Dimension Date'[Calendar Year] DESC
```

The DEFINE MEASURE statement creates an area where you can add multiple measures. Each measure must have the full notation of 'tablename'[measure name] = preceding the DAX expression. The code used in the expression is the same as what you use when creating a calculated measure directly in the data model. The measure name must include a table.

The measure can then be used in the EVALUATE query with the results visible in the results tab when it is run (Figure 8-7). Syntax error messages appear in the Output tab of the Results panel if there are errors in the code.

Results

Calendar Year	My Measure
2016	29518
2015	71828
2014	65957
2013	60678
	284

Figure 8-7. *Output of code from Listing 8-11 using DEFINE MEASURE*

Another element of running queries in a client tool such as DAX Studio is that you can add an ORDER BY statement to the query; this is also shown in Listing 8-11.

These are the basics for getting queries to run in DAX Studio. Additional features are useful for optimizing queries and we look at these in more detail later in this chapter.

In addition to having a larger area for the query editor and a host of tools that are useful when you're editing longer DAX queries, you can save and store your queries as separate files. Doing so can help you back up and provide source control over individual queries.

The Format Query (F6) feature allows you to tidy queries by adding tabs and carriage returns at key places to help improve query readability.

You can run multiple queries at the same time, as shown here:

```
DEFINE MEASURE 'Fact Sale'[Right Now] = FORMAT(NOW(),"hh:mm:ss.ms")

EVALUATE
 ROW("Q1",[Right Now])

EVALUATE
 ROW("Q3",[Right Now])

EVALUATE
 SAMPLE(10000,'Fact Sale','Fact Sale'[Invoice Date Key])
```

When you run this query, it returns three datasets to the Results panel.

You can view the results for each EVALUATE statement by clicking the numbered vertical tabs. The result of the third and final EVALUATE query from the batch is shown in Figure 8-8. Note that two of the queries used the calculated measure from the DEFINE MEASURE section. The final query uses the SAMPLE function to perform a random filter over 1,000 rows from the 'Fact Sale' table.

Sale Key	City Key	Customer Key	Bill To Customer Key
227997	99051	0	0
228205	44781	385	202
228034	97570	0	0
227988	87114	397	202
228217	102671	0	0
228080	86258	0	0
228209	77447	0	0
228009	96334	291	202

Figure 8-8. *The last result set from three queries run simultanously*

SSMS (SQL Server Management Studio)

Another tool to help develop and debug your DAX calculations is Microsoft SSMS. If you are working with Microsoft SQL databases, you probably already have this tool. If you don't, a quick internet search will point you to the correct download location for this on the Microsoft site. It is a large download because it contains features useful for working with the entire Microsoft SQL suite of products and not just DAX models.

Just like DAX Studio, you can use SSMS to connect to DAX data models such as Power BI Desktop and SSAS Tabular. However, to connect to Power BI Desktop, you need to know the port number on your local machine being used by the instance you want to connect to. This number changes each time you open Power BI Desktop. DAX Studio shows the port number once you are connected on the Status bar in the format of *localhost:nnnn* where *nnnn* is the port number. So, if you have DAX Studio, you can use this as a quick way to identify the port number you need if you also want to use SSMS to connect to your model.

Note   You can start Power BI Desktop on a specific port by navigating to the folder where the Desktop executable is located and running this command from the console:

```
PBIDesktop /diagnosticsport:XXXX
```

Another way to find the current port number is via the Resource Monitor (Figure 8-9) for your operating system (or a similar tool); look for any listening network ports using the msmdsrv.exe executable. This is easier if you only have one instance of Power BI Desktop open.

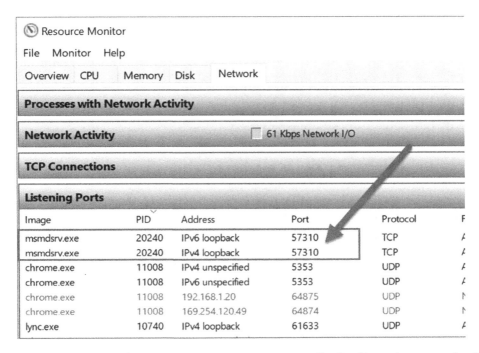

Figure 8-9. *Using Windows 10 Resource Monitor to find a listening port for Power BI Desktop*

In this case, the Resource Monitor is showing that port number 57310 is a listening port by msmdsrv.exe, so you can use it when connecting from SSMS using the properties in Figure 8-10.

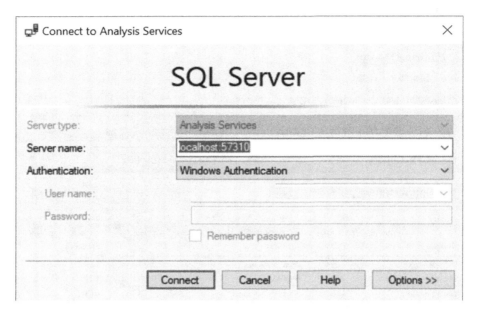

Figure 8-10. *Example connect dialog for SSMS for connecting to the Power BI Desktop*

Note   The Server Type needs to be set to Analysis Services. This might not be the default value of this property.

Once connected, DAX queries must conform to the same requirements as those you use for DAX Studio. These include that you need to use the EVALUATE keyword for every query and that all results should be in a DAX table.

Otherwise SSMS provides IntelliSense and a host of other features to help you build DAX calculations. All the query examples used in this chapter to demonstrate DAX Studio work in SSMS except for multiple EVALUATE statements in the same batch.

Other Client Tools

Once you know the server and port number for the DAX engine you wish to query, you can use any tool that has the ability connect to an instance of SSAS Tabular. You can even connect Power BI Desktop to another instance of Power BI Desktop. This is probably not so useful for developing DAX queries, but doing so can help optimize your data model by allowing DMV queries to be issued with the results plotted using the fantastic array of visuals in Power BI.

Optimizing

Optimization is a broad and potentially advanced topic—enough to justify a book on its own, in fact. In this section, I try to cover some of the more useful techniques that you can used to improve the performance of your model. I have split the suggestions into two sections. The first section is for optimizing data models and the second focuses on optimizing DAX queries and also covers how to take advantage of tools such as DAX Studio and SSMS to improve performance.

Optimizing Data Models

Let's begin by removing unused data.

Removing Unused Data

The first tip is to make sure you only include data you need in the model. The larger the model, the more work calculations you need to do, and surplus data, even if it's not used in any calculation, can still have a detrimental effect on compression and performance of the model.

Although it might be initially useful to import every row and every column from various tables in your source system for data exploration purposes, make sure you allow time to review and remove unused columns and rows from your data model before you publish your report to a production environment.

Highly unique data sitting in unused columns can have a negative effect on the compression used by the DAX engine to store the data. Good columns to target to find such unused data are IDENTITY or Autonumber columns that are useful to have in an OLTP source system but that often have little value in an OLAP data model. Unless you need to display values from these columns in your report, removing them can help considerably reduce the size of your data model.

You can use a free tool called Power BI Helper to help identify columns that aren't being used in a model. You should remove any column that it highlights from the data model and only add them back if you need them for the report. You can download the Power BI Helper from `http://radacad.com/power-bi-helper`.

If you are comfortable using DMV, then issuing the following DMV query against your data model using DAX Studio or SSMS, once sorted, can provide you with useful insight into which objects (columns) in your model are using the most memory.

```
Select * from $System.discover_object_memory_usage
```

These columns are good ones to check to see if they can be removed.

Removing unused columns is one technique, but also consider the number of rows you import into your model. If the data model only needs to support a certain range of time, then consider importing just enough data to cover the period required.

Also consider compressing data as part of the import process. If a dataset from a source system has many rows per day, such as the kind you might find in some IOT Sensor data, group the data before or during the import. It may be possible to satisfy reporting requirements using data that has been presummarized down to a single row per day that also contains a value for the COUNT, MIN, MAX, and AVERAGE of the rows that have been compressed.

These suggestions are helpful to reduce the overall size of the data model. If you try these out using a Power BI Desktop model, make a note of the file size of the saved PBIX file before and after the changes. I have seen instances where the file size was reduced from over 800 MB down to less than 40 MB without breaking any visuals in the report. This was mainly a combination of running the suggested DMV against the data model, identifying column objects using the most memory, and removing them from the model. In this case, most columns removed were IDENTITY columns sitting in tables with several hundreds of millions of rows.

Note   Hiding a column in the model does not affect performance. Remove unused columns from the data import logic.

Creating Summary Tables

Another technique that doesn't address the overall size of the model is to create additional summary tables inside your model. You can generate these tables either in DAX or as part of the import process in the query editor.

If you have a large table with many rows that you need kept in its original form to meet some reporting requirements, consider creating summary tables you can use in visuals and calculations that don't need to drill down to the finest detail of the original table.

Take the example of the 'Fact Sale' table in the WideWorldImportersDW dataset. This contains 228,265 rows with detail on every sale.

Creating a summary version of this table by Calendar Month and Customer in this dataset would reduce this down to 5,239 rows. Such a summary table can support any visual or calculation that requires this level of detail and it is 1/50th of the size. Performance improvements using this approach are amplified when you are creating reports using tables that have many hundreds of millions of rows.

You can create dozens of summary tables from a single large table to provide a highly customized level of performance tuning. In this event, I recommend a good naming convention to help you keep track.

Precalculating Data

Consider using calculated columns instead of calculated measures for metrics that are not required to be dynamic. An example of this might be a ranking measure that determines who are the best customers or sales people based on data in a large table. It's possible to perform these types of calculations as calculated measures or calculated columns. However, if you don't need to dynamically recalculate the metric to respect changing slicer or filter selections, then depending on the size of the data, using a calculated column will probably improve the performance of your model.

Splitting Up Unique Columns

If you have a column in your dataset that has highly unique values, such as a DateTime column that includes detail from year to millisecond, or an IDENTITY or Autonumber column, then consider splitting it into multiple columns.

The DateTime example is straightforward in that it can be split neatly into one column that carries a value for day, while other columns carry values for hours, minutes, or seconds. If you don't need to report to a level of detail finer than day, then remove this information from the data model altogether.

Look for other opportunities to split columns where the data can have multiple meanings. These might include values that represent the make and model of a car, or address data that includes a level of detail that isn't useful for the reports, such as suburb or street.

Simplifying Table Structure

DAX data models can provide accurate calculations using a table structure that mirrors the table structure of your source system. This includes calculations that span many generations of a set of table relationships. However, your model will probably perform much faster if you flatten your data structures to reduce the number of relationships that need to be used per calculation.

A good model to follow is to organize your data into fact and dimension tables. This is commonly known as a star schema and it has been proven to work well for reporting and analytic workloads.

Fact Tables

Start by identifying entities in your data that you would like to count, sum, and generally visualize over time. These entities should form the basis of fact tables in your model. Each row in a fact table should be an instance, activity, event, or transaction that the fact table represents. If you have 100 sales, then ideally a 'Fact Sales' table should have 100 rows.

Do not try and mix multiple facts in a single table. If you need to report and analyze sales and payments, avoid the temptation of combining both sets of data in the same fact table. Have a fact table for sales and another fact table for payments.

Avoid creating relationships between fact tables. Many useful calculations can be performed over unrelated fact tables that share a common dimension table.

Once you have identified the fact tables you need, bring as many columns of data from other source system tables into the fact table as you can to satisfy known reporting needs, but be careful to still retain the *1 row = 1 fact* rule. If you can make the fact table 100-percent self-sufficient, you will get the best performance. Only make exceptions for columns that might be useful in a dimension table that are connected to other fact tables—for example, rather than replicate a month, quarter, and year table in every fact table, move these to a dimension table.

Try to avoid data models that involve calculations that use data using relationships that span three or more levels.

Dimension Tables

Think of dimension tables as filter/slicer tables that are only useful for slicing and dicing your fact tables. They generally don't contain data you intend to count, sum, or use in calculations. If they do, consider if this data should be used in a fact table instead.

Dimension tables are most useful for providing a single place for filter selections to apply across relationships to multiple fact tables.

The first dimension table you add to any model will probably be a date table. If you have multiple fact tables, then this likely has a relationship to each of these with a value that represents a single day. Any filter/slicer selection made to any column in this table propagates down through to filter any connected fact table.

If you have a dimension table that is only related to one table, consider combining the two into a single table.

Optimizing DAX Queries

It's possible to write a query many ways and get the same result. Five people may produce five different approaches that each have unique performance characteristics. In this section, I cover some of the more useful tips and techniques for making DAX calculations more efficient. Some of these are enough to help you solve a performance issue you may encounter. Other solutions are a mix of improvements made to DAX calculations and enhancements to the underlying data structures.

In this section, I also show you how you can use tools such as DAX Studio and SQL Server Profiler to help optimize your DAX queries.

Storage Engine vs. Formula Engine

DAX queries use two engines when running calculations: the storage engine (SE) and the formula engine (FE).

The *storage engine* is the faster of the two and has the primary objective of retrieving all the data it needs to compute the query from the in-memory data stores. The storage engine is multithreaded, so it can complete multiple operations in parallel. The storage engine can perform simple calculations while retrieving data, but these need to be calculations that don't rely on any other storage engine task that might be happening in parallel as part of the same query.

Results from storage engine activity can be stored in the cache to help speed up subsequent queries. In the case where you have a calculated measure that is executed many times by the same visual to produce a value for each data point, this caching can play a role in helping speed up all the calculations used in a visual.

The *formula engine* is single-threaded and capable of more sophisticated logic including calculations in which the order of the steps involved is important to the result. Results from formula engine activity are not cached so they must be recalculated each time.

When a query is passed to the DAX engine, the storage engine starts by retrieving relevant data and the formula engine completes the job.

A good approach to use to optimize a slow DAX query is to try and understand which components of the query are being completed by the storage engine and which are being completed by the formula engine; then see if some of the work being handled by the formula engine can be moved to the storage engine.

A quick way to tell how much of your query is using the storage engine instead of the formula engine is by using DAX Studio. There is an option on the ribbon (Figure 8-11) to enable server timings as well as the query plan.

Figure 8-11. *Options on the DAX Studio ribbon for turning on additional debug options*

With these options enabled, any individual query run in DAX Studio now generates additional information that shows how long the engine took to complete the query. This also shows the breakdown of how much of the overall time was used by the storage engine instead of the formula engine (Figure 8-12).

Total	**SE CPU**
15 ms	0 ms
	x0.0

FE	SE
12 ms	3 ms
80.0%	20.0%

SE Queries	**SE Cache**
2	0
	0.0%

Figure 8-12. *Example output of server timings for a query run in DAX Studio*

Figure 8-12 show an example of a query that took a total of 15 milliseconds to complete. The first 3 milliseconds were spent by the storage engine, which ran two parallel processes to collect all the data required. When the storage engine finished, the formula engine took over and needed 12 milliseconds to finish the query. The formula engine was active for 80 percent of this query, and the storage engine did not benefit from the storage engine cache.

This information doesn't explain which parts of your DAX query were handled by the storage engine vs. the formula engine, but it does give you a baseline for understanding how other versions of the same query behave.

When optimizing, it is important to clear the cache each time you run an updated version of your query. If you don't, you may get the impression that your newer version is better because the overall time is faster, where in fact it could be taking advantage of cached data from a previous query. This is easy to do in DAX Studio by setting an option on the Run button to clear the cache each time.

To understand which components of a query are being handled by the storage engine or the formula engine, you need to dive down into a query plan generated by a trace over the server. Just as you can run traces on a Microsoft SQL server to analyze T-SQL queries, SSAS engines can provide the same ability to capture events that show various subcomponents of individual queries. Because Power BI Desktop uses an SSAS engine, you can use tools such as SQL Server Profiler to capture interesting events while a query is run to show exactly how the DAX engine went about the task of completing the query. Certain events show as being activities completed by the storage engine, while others show up as belonging to the formula engine.

This is quite an advanced topic, so apart from showing you how to run a trace over a DAX query later in this chapter and from showing you that it is this underlying trace data that DAX Studio uses to obtain the server timings, I do not dig deeper. DAX Studio shows both the physical and logical query plans used by the query in a Query Plan tab along with some pseudo-T-SQL generated by the underlying trace to show the work of each storage engine subquery.

Filtering Early and Appropriately

If you are tasked with producing a handwritten list of all the words in a dictionary that began with the letter X, one approach is to start by making a list of every word in the dictionary from A to Z and then cross out all the words that don't begin with the X. Or you and five friends can each take a copy of the same dictionary and start writing words from different pages that have words starting with the letter X. The second approach is much faster due to the fact that less work is required and that it is shared.

You can apply the same approach to DAX calculations. The earlier you can apply a filter without affecting the result, the bigger the impact doing so has on the overall time the calculation takes.

The two DAX calculations in Listings 8-12 and 8-13 demonstrate this nicely. Both calculations return identical results. The first query shows a sum of a quantity column along with a year-to-date calculation. The data is being filtered to a specific sales territory and month of the year.

Listing 8-12. A DAX Query Using FILTER After SUMMARIZE

```
EVALUATE
    FILTER(
        SUMMARIZE(
            'Fact Sale',
            'Dimension Date'[Date],
            'Dimension Date'[Calendar Month Number],
            'Dimension City'[Sales Territory],
            "SUM Qty", SUM('Fact Sale'[Quantity]),
            "YTD Qty", TOTALYTD(SUM('Fact Sale'[Quantity]),'Dimension
            Date'[Date])
            )
        ,
        'Dimension City' [Sales Territory] = "Southeast"
        && 'Dimension Date'[Calendar Month Number] =11
        )
    ORDER BY
        'Dimension Date'[Date]
```

This query produces 77 rows along with the output in Figure 8-13 in the Server Timing window in DAX Studio.

Total	SE CPU	Line	Subclass	Duration	CPU	Rows	KB
301 ms	32 ms	2	Scan	3	16	19,032	298
	x1.5	4	Scan	1	0	1,464	12
FE	SE	6	Scan	12	16	8,998	36
280 ms	21 ms	8	Scan	0	0	1,464	12
93.0%	7.0%	10	Scan	5	0	19,032	298

SE Queries	SE Cache
5	0
	0.0%

Figure 8-13. *The server timing output for the code in Listing 8-12 using DAX Studio*

The query took 301 milliseconds in total and spent a considerable amount of that time in the formula engine. This query was executed multiple times over a cold cache and this timing is typical.

Listing 8-13 shows another query that produces the same output.

Listing 8-13. Alternate Version of Listing 8-12 Applying a Filter Earlier in the Logic

```
EVALUATE

VAR PreFilteredTable =
    CALCULATETABLE(
        SUMMARIZE(
            'Fact Sale',
            'Dimension Date'[Date],
            'Dimension Date'[Calendar Month Number],
            'Dimension City'[Sales Territory]
        ),
        'Dimension City'[Sales Territory] = "Southeast",
        'Dimension Date'[Calendar Month Number] =11
    )
```

```
RETURN
    ADDCOLUMNS(
            PreFilteredTable,
            "SUM Qty", CALCULATE(SUM('Fact Sale'[Quantity])),
            "YTD Qty", TOTALYTD(SUM('Fact Sale'[Quantity]),'Dimension
            Date'[Date])
            )

ORDER BY
    'Dimension Date'[Date]
```

This version also produces 77 rows and the output of the Server Timing window is shown in Figure 8-14.

Total	SE CPU	Line	Subclass	Duration	CPU	Rows	KB
44 ms	16 ms	2	Scan	4	0	1,464	23
	x1.1	4	Scan	2	0	1,464	12
FE	SE	6	Scan	1	0	1,464	12
29 ms	15 ms	8	Scan	1	16	1,464	12
65.9%	34.1%	10	Scan	7	0	1,464	23

SE Queries	SE Cache
5	0
	0.0%

Figure 8-14. *The server timing output for faster code in Listing 8-13 using DAX Studio*

The updated version took a total of 44 milliseconds. Both versions used five storage engine queries to retrieve the data, but in the updated query, filtering took place before the results were passed to the SUM and TOTALYTD functions. The storage engine returned smaller batches to the formula engine. None of the storage engine queries had more than 1,464 rows, whereas the previous query had storage engine results of over 19,000 rows passed to the formula engine for extra processing.

Hopefully this example shows you how you might take advantage of some very helpful information that is available in DAX Studio. You do not need to run every query through this process, but isolating slower calculations to be tuned and optimized in DAX Studio is an effective use of time.

Another useful tip regarding filters in DAX, particularly the use of filter functions, is that that these functions, such as the ALL function, can often operate over tables or columns. If you can achieve the same result by using a specific column rather than an entire table, you will probably enjoy some performance gains. Passing an entire table can cause unnecessary work, especially if passing just a table column can achieve the same result.

SQL Server Profiler

To run a trace using Profiler, start Profiler directly or via SSMS. Make a connection to an instance of Analysis Services using a server name that your DAX engine is listening on. If you are connecting to an instance of Power BI Desktop, you need to know the port number being used by that instance. This changes each time you stop and start Power BI Desktop. DAX Studio shows the port number in use on the Status bar, otherwise you can get this information using process monitoring tools such as Resource Monitor or the Windows Sysinternals version of Process Explorer.

These are some useful events to enable for each trace:

- Query Begin
- DAX Query Plan
- Query Subcube
- VertiPaq SE Query End

Start the trace, then run your DAX query, then stop or pause the trace at a point where you can review the results.

The diagram in Figure 8-15 shows events captured by Profiler when it was used to trace the second version of the previous query. More detail is available when you click on individual rows in the trace window.

EventClass	EventSubclass	Duration	TextData
Query Begin	3 - DAXQuery		EVALUATE RO
DAX Query Plan	1 - DAX VertiPaq Logical Plan		AddColumns:
DAX Query Plan	2 - DAX VertiPaq Physical Plan		AddColumns:
Query End	3 - DAXQuery	64	EVALUATE RO
Query Begin	3 - DAXQuery		EVALUATE V.
DAX Query Plan	1 - DAX VertiPaq Logical Plan		Order: RelL
VertiPaq SE Query End	10 - Internal VertiPaq Scan	4	SET DC_KIND
VertiPaq SE Query End	0 - VertiPaq Scan	4	SET DC_KIND
VertiPaq SE Query End	10 - Internal VertiPaq Scan	2	SET DC_KIND
VertiPaq SE Query End	0 - VertiPaq Scan	3	SET DC_KIND
VertiPaq SE Query End	10 - Internal VertiPaq Scan	0	SET DC_KIND
VertiPaq SE Query End	0 - VertiPaq Scan	0	SET DC_KIND
VertiPaq SE Query End	10 - Internal VertiPaq Scan	1	SET DC_KIND
VertiPaq SE Query End	0 - VertiPaq Scan	1	SET DC_KIND
VertiPaq SE Query End	10 - Internal VertiPaq Scan	9	SET DC_KIND
VertiPaq SE Query End	0 - VertiPaq Scan	11	SET DC_KIND
DAX Query Plan	2 - DAX VertiPaq Physical Plan		PartitionIn
Query End	3 - DAXQuery	49	EVALUATE V.

Figure 8-15. *Sample of output for code in Listing 8-12 using Profiler*

It's also possible to enable tracing in Power BI Desktop by turning on the Enable Tracing option in the settings under diagnostics (Figure 8-16). This creates *.trc files that you can open and analyze using Profiler. The *.trc files are stored in the folder shown in the same setting window.

Figure 8-16. *Option in Power BI Desktop settings that allows trace files to be generated*

There often plenty of options available to help your queries run faster. Some involve you modifying the data model itself, and you can achieve others through changes to DAX calculations. In general, if you remember that calculations using the storage engine are good, and that it is also preferable to have data models that are small and simple, you are probably going to end up with a well-optimized model (and happy end users).

Remember to also make effective use of online material. DAX has been around since 2009 and there is a fantastic community of users that share interesting ways to approach common problems.

CHAPTER 9

Practical DAX

This chapter focuses on building a model using DAX and without using external data sources. In this chapter, you build your own dataset. All data is generated using step-by-step examples that showcase various tips and techniques you may find useful.

The final model has a structure resembling what you might use for a small organization with tables for sales, dates, and products. Because the number of rows for each table is automatically generated, you can use the final model for optimizing DAX calculations over a large dataset. By changing a few numbers here and there, you can generate a sales table with 5, 50, or 500 million rows of data with which to test different approaches to optimizing calculations.

These step-by-step exercises rely on you using a version of DAX that can create calculated tables. This includes Power BI Desktop and SSAS Tabular but not PowerPivot for Excel 2016 (or earlier).

Creating a Numbers Table

The first step is to create a numbers table, in this case, a single-column table with a sequence of numbers starting at 0 and running to 1,000. You will be able to use this utility table to join to other tables using range-based conditions, such as when the value in this table is less than 5. In this exercise, the numbers table is used to multiply a single row from a table into many rows.

To create a numbers table you have several options. Before the GENERATESERIES function became available, the only way to create this type of table was to take advantage of the CALENDAR function. Not all versions of DAX have the GENERATESERIES function. Dates in DAX are stored as integer values, so you can use the CALENDAR function to generate rows of dates that you can then convert back to integer values. In Power BI Desktop, the date December 30, 1899, is stored as 0. The value of 1 belongs to the December 31, 1899, and 2 is January 1, 1900. Interestingly, the value of –1 belongs to

© Philip Seamark 2018
P. Seamark, *Beginning DAX with Power BI*, https://doi.org/10.1007/978-1-4842-3477-8_9

December 29, 1899, so if you need to represent data back in the 1600s with DAX, you can, but you need to use numbers that are in the vicinity of –100,000.

Here is how you can create a numbers calculated table (see Figure 9-1) using the CALENDAR function:

```
Numbers =
    SELECTCOLUMNS(
        CALENDAR(0,1000),
        "N", INT ( [Date] )
    )
```

To achieve the same result using the GENERATESERIES function, the code you use to create the calculated table is

```
Numbers =
    SELECTCOLUMNS(
        GENERATESERIES( 0,1000 ),
        "N", [Value]
    )
```

Both examples produce a table with 1,001 rows called Numbers with a single column called N. The SELECTCOLUMNS function is used in both cases to rename the column to N. The INT function in the example using the CALENDAR function is used to convert the date output to its underlying number.

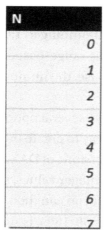

Figure 9-1. *A sample of the Numbers table*

Adding a Date Table

The next step is to add a date table to the data model. You can use the DAX in Listing 9-1 to generate this.

Listing 9-1. Calculated Table to Create a Simple Date Table

```
Dates =
VAR BaseTable =
    CALENDAR (  DATE(2016, 1, 1), TODAY () )

VAR AddYears =
    ADDCOLUMNS (
        BaseTable,
        "Year", YEAR ( [Date] )
        )

VAR AddMonths =
    ADDCOLUMNS (
        AddYears,
        "Month", DATE(YEAR([Date]),MONTH([Date]),1),
        "Month label", FORMAT ( [Date], "MMM YY" )
    )

VAR AddDay =
    ADDCOLUMNS (
        AddMonths,
        "Day Label", FORMAT ( [Date], "DDD d MMM YY" )
        )
RETURN
    AddDay
```

This statement begins with the single column output of the CALENDAR function, which is then assigned to the BaseTable variable. The virtual table has one row per day starting on January 1, 2016, and running sequentially through until the current date.

You can use additional variables to incrementally add columns throughout the calculation. In this code, the AddYears variable takes the table represented by the BaseTable variable and appends a column called "Year" using the YEAR function to extract a value from the [Date] column.

233

The AddMonths variable uses ADDCOLUMNS to append two additional columns. These allow rows to be grouped into calendar months. The "Month" column uses the DateTime datatype and converts each date value back to be the first date for the month it represents. The "Month label" column uses the FORMAT function to convert dates to text. The format string notation of MMM YY means dates such as July 4, 2018, are converted to a text value of "Jul 18". You need to set the Sort by Column property of the "Month label" column to use the "Month" column, otherwise values in this column appear alphabetically when used in visuals.

The final AddDay variable uses the ADDCOLUMNS function to add a column called "Day Label". This uses the FORMAT function to convert the date from each row into a text value using the format string, "DDD d MMM YY". This format string converts July 4, 2018, to "Wed 4 Jul 18". This text-based column needs to have its Sort by Column property set to the "Date" column to ensure sensible sorting when used in visuals. Figure 9-2 shows the first few rows generated by the date table.

Date	Year	Month	Month label	Day Label
1/01/2016 12:00:00 AM	2016	1/01/2016 12:00:00 AM	Jan 16	Fri 1 Jan 16
2/01/2016 12:00:00 AM	2016	1/01/2016 12:00:00 AM	Jan 16	Sat 2 Jan 16
3/01/2016 12:00:00 AM	2016	1/01/2016 12:00:00 AM	Jan 16	Sun 3 Jan 16
4/01/2016 12:00:00 AM	2016	1/01/2016 12:00:00 AM	Jan 16	Mon 4 Jan 16
5/01/2016 12:00:00 AM	2016	1/01/2016 12:00:00 AM	Jan 16	Tue 5 Jan 16
6/01/2016 12:00:00 AM	2016	1/01/2016 12:00:00 AM	Jan 16	Wed 6 Jan 16
7/01/2016 12:00:00 AM	2016	1/01/2016 12:00:00 AM	Jan 16	Thu 7 Jan 16
8/01/2016 12:00:00 AM	2016	1/01/2016 12:00:00 AM	Jan 16	Fri 8 Jan 16

Figure 9-2. *Sample output of the date table created by Listing 9-1*

Creating the Sales Table

The next table you need to create is a sales table. The objective is to generate a table that shows sales over a timespan that you can use to build a sales report. This section covers each step you need to use showing the DAX and explaining the code.

The first task is to generate a random number of sales for each day. The DAX for this is shown in Listing 9-2.

Listing 9-2. Excerpt from a Calculated Table to Create a Sales Table

```
Sales =
VAR FirstTable =
    SELECTCOLUMNS(
        FILTER (
            GENERATE (
            Dates,
            Numbers
                ),
            [N] < RANDBETWEEN ( 2, 7 )
        ),
        "Date",[Date])
```

The output of this statement is a table with a single column of dates. Each date occurs anywhere from two to six times and effectively creates a placeholder row that you can use to add columns. The CROSSJOIN and FILTER functions are used to combine the existing date and numbers tables you created earlier. Without the FILTER function, the CROSSJOIN matches each row from the date table to all 1,001 rows in the numbers table. It's unlikely that a real sales table would have exactly 1,001 sales per day, so the FILTER function uses the RANDBETWEEN function to create a more randomized number of sales per day. In this case, each day has somewhere between two and seven sales transactions.

You can set the lower boundary to zero so you have some days with no sales, and you can increase the upper bound as one way of generating a larger set of sales.

You can use the SELECTCOLUMNS function to ensure that only the columns you need from the date and numbers tables flow through to the output, which is assigned as a virtual table to the FirstTable variable in Listing 9-2.

The next task is to add a column called "Product" (Listing 9-3) that carries a value to represent the product that was sold.

Listing 9-3. Adding a Product Column to Listing 9-2

```
VAR AddPRoduct =
ADDCOLUMNS (
    FirstTable,
    "Product",
```

```
    VAR myMake =
        RANDBETWEEN ( 65, 90 )
    VAR myModel =
        FORMAT ( RANDBETWEEN ( 1, 5 ), "-#" )
    RETURN
        REPT ( UNICHAR ( myMake ), 3 ) & myModel
)
```

This code starts by using the FirstTable variable as input for the ADDCOLUMNS function. A nested variable called Make begins a new level of variable scope. This scope is separated from the layer being used by the FirstTable and AddProduct variables, and the RETURN function is used to drop a single value back as the third parameter of the ADDCOLUMNS function. You can achieve the same result without using a nested variable, but it would be harder to read.

The myMake variable uses the RANDBETWEEN function to generate a random number between 65 and 90. These numbers are used because they represent the range on the ASCII table for the uppercase letters of the alphabet. The value of 65 represents the letter "A," 66 is "B," and this continues to 90, which is the uppercase "Z." The random number is stored in the myMake variable.

The myModel variable uses the RANDBETWEEN function to generate a random number between 1 and 5. This is then converted to text using the FORMAT function with a format string of "-#". The hyphen in the format string is preserved while the hashtag placeholder tells FORMAT where the number should go. This creates a string such as "-1" or "-5".

The final RETURN statement concatenates the myMake and myModel variables together to produce a text value that looks like a product code. The number stored in myMake is first converted to an uppercase letter between A and Z using the UNICHAR function. The output of the UNICHAR function is then repeated three times using the REPT function. This turns a value of "A" into "AAA", and when combined with the text in the myMake variable, it produces a result such as "AAA-4" or "DDD-1".

At this point the output of the calculated table looks like Figure 9-3.

Date	Product
1/01/2016 12:00:00 AM	PPP-5
1/01/2016 12:00:00 AM	CCC-4
1/01/2016 12:00:00 AM	LLL-5
1/01/2016 12:00:00 AM	III-4
1/01/2016 12:00:00 AM	WWW-5
2/01/2016 12:00:00 AM	YYY-5
2/01/2016 12:00:00 AM	VVV-2

Figure 9-3. *Sample output of the sales table including Listing 9-3*

UNICHAR is a great function that you can use to add KPI indicators, such as trend arrows or emoji pictures, to reports to emphasize a point.

The next column that is added represents a quantity sold (Listing 9-4). This builds on the existing code by passing the AddProduct variable as the first parameter of the ADDCOLUMNS function.

Listing 9-4. Adding the Qty Column to the Sales Table

```
VAR AddQuantity =
    ADDCOLUMNS (
        AddProduct,
        "Qty",
            VAR Q = RANDBETWEEN ( 1, 1000 )
            RETURN 5 - INT ( LOG ( Q, 4 ) )
    )
```

The ADDCOLUMNS function appends a single column called "Qty". A nested variable scope begins here in the third parameter. The intention is not only to generate a random number between 1 and 5, but to make sure the distribution of random values is not even. In a real-life sales table, it's likely that most transactions will involve a quantity of 1. The next most common is 2 and the number of purchases reduces as the quantity increases.

The LOG function used here applies a logarithmic distribution for quantities between 1 and 5. When this code is added to the calculated table statement and plotted on a visual using the count of each value in the Qty column (Figure 9-4), it shows most transactions are single-quantity transactions. The number of transactions with a quantity of 2 is nearly half that of 1, whereas transactions using a quantity of 3 drop off by more than half.

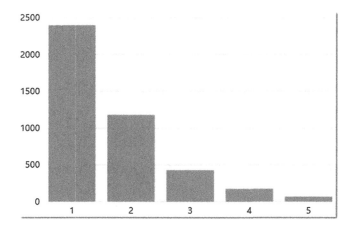

Figure 9-4. *Chart showing distribution of quantity values*

This is the desired effect of the calculation and you can tweak it further to suit your needs.

The output of the sales calculated table using the code so far is similar to Figure 9-5.

Date	Product	Qty
1/01/2016 12:00:00 AM	BBB-5	1
1/01/2016 12:00:00 AM	GGG-3	1
1/01/2016 12:00:00 AM	XXX-2	1
1/01/2016 12:00:00 AM	BBB-3	2
2/01/2016 12:00:00 AM	RRR-5	1
2/01/2016 12:00:00 AM	TTT-3	1
2/01/2016 12:00:00 AM	AAA-4	3
2/01/2016 12:00:00 AM	TTT-5	4

Figure 9-5. *Sample of the sales table including the QTY column*

The next column added is for Price. The code for this column is shown in Listing 9-5.

Listing 9-5. Adding a Price Column to the Sales Table

```
VAR AddPrice =
    ADDCOLUMNS (
        AddQuantity,
        "Price", DIVIDE (
                    RANDBETWEEN ( 1, 1e5 ),
                    100 )
                    )
```

The ADDCOLUMNS function appends a column to the table stored in the AddQuantity variable. The new column is called "Price" and the formula uses the RANDBETWEEN function to generate a number between 1 and 100,000. The notation of 1e5 shows that scientific notation can be used as an option. The value of 1e5 is shorthand for 1×10^5 (or 1×10 to the power of 5), which equals 100,000.

Once this random number is generated, the DIVIDE function is used to convert this to a value that represents a value of 1,000 or less, with two decimal places. A feature of the DIVIDE function is that it handles "divide by zero" errors better than the / operator; however, the / operator performs better and you should give it preference if there is no chance of a "divide by zero" error. Here is an example:

```
RANDBETWEEN( 1, 1e5) / 100
```

In Listing 9-5, there is no attempt to align prices across product codes. If having price values consistent across products is important for your testing, you can shift the logic shown here to generate a random price to a product calculated table and add it to the sales table via a relationship.

The final column that needs to be added combines [Price] and [Qty] to create a [Total] column. This demonstrates that a new column can use values derived from earlier parts of the same code.

Listing 9-6 is the full version of the code that creates the sales calculated table.

Listing 9-6. The Complete Calculated Table for the Sales Table

```
Sales =
VAR FirstTable =
SELECTCOLUMNS(
    FILTER (
        GENERATE (
            Dates,
            Numbers
        ),
        [N] < RANDBETWEEN ( 2, 7 )
    ),
    "Date",[Date])

VAR AddPRoduct =
    ADDCOLUMNS (
        FirstTable,
        "Product",
        VAR Make =
            RANDBETWEEN ( 65, 90 )
        VAR myModel =
            FORMAT ( RANDBETWEEN ( 1, 5 ), "-#" )
        RETURN
            REPT ( UNICHAR ( MAke ), 3 ) & myModel
    )

VAR AddQuantity =
    ADDCOLUMNS (
        AddPRoduct,
        "Qty",
            VAR Q = RANDBETWEEN ( 1, 1000 )
            RETURN 5 - INT ( LOG ( Q, 4 ) )
    )

VAR AddPrice =
    ADDCOLUMNS (
        AddQuantity,
```

```
      "Price", DIVIDE (
                RANDBETWEEN ( 1, 1e5 ),
                100 )
                )
VAR AddTotal =
    ADDCOLUMNS ( AddPrice, "Total", [Price] * [Qty] )
RETURN
    AddTotal
```

Figure 9-6 shows a sample of the table that uses this code.

Date	Product	Qty	Price	Total
1/01/2016 12:00:00 AM	RRR-5	1	836.66	836.66
1/01/2016 12:00:00 AM	CCC-3	1	216.53	216.53
1/01/2016 12:00:00 AM	OOO-1	1	893.3	893.3
1/01/2016 12:00:00 AM	EEE-5	1	743.85	743.85
2/01/2016 12:00:00 AM	BBB-2	1	587.4	587.4
2/01/2016 12:00:00 AM	SSS-1	2	790.76	1581.52
2/01/2016 12:00:00 AM	ZZZ-3	2	534.35	1068.7
2/01/2016 12:00:00 AM	GGG-5	4	75.68	302.72
2/01/2016 12:00:00 AM	OOO-5	1	433.03	433.03

Figure 9-6. *Sample output of the sales table including the Total column*

The table now has five columns and potentially hundreds of millions of rows. You can control the number of rows with the calculation that is used to generate the numbers table along with the FILTER criteria for the FirstTable variable. Bear in mind that each change you make to the any part of any of the calculated tables causes the data to recalculate for every table in the model. This generates new values each time but could take a long time to calculate if you have it configured to generate extremely large datasets.

The sales table can now be related to the date table using the [Date] column in each table.

Optimizing Sales

The data model now has a detailed sales table with rows that represent individual transactions. This table could grow to be very large. The sales table could have the same product appear in multiple rows for the same day. This is more likely to happen if the code to generate the sales table is deliberately set to generate a very large table. When it comes to building reports using tables in the dataset, not all report visuals need to use the finest level of detail available in the larger tables.

For instance, if you have a visual that only needs to show the sum of the Qty column by day and never by product, using the sales table can be quite an inefficient approach.

An alternative is to generate a summarized version of the sales table grouped by day, but have no ability to slice and dice by product.

The code to generate a calculated table for this is shown in Listing 9-7.

Listing 9-7. Summary of the Calculated Table of the Sales Table in Listing 9-6

```
Daily Summary =
    SUMMARIZECOLUMNS(
        'Sales'[Date],
        "Sum of Items Sold", SUM('Sales'[Qty]),
        "Sum of Revenue", SUM('Sales'[Total])
    )
```

This creates a three-column table that summarizes multiple lines per 'Sales'[Date] down to a single line while generating a value for each row using the SUM function to provide aggregate totals over the Qty and Total columns.

A visual using this table is going to be faster than one using the sales table, especially if the sales table ends up with a very large number of rows.

A downside of creating summary tables such as this is that they increase the memory footprint of the overall data model and make the data refresh process take longer. The other obvious downside is that any visual using the summary table can only be sliced and diced by the grain in the table, which in this case is [Date] and not [Product].

If you have a report that is slow to interact with and does not use summary tables, this is a strategy well worth considering.

Calculated Columns vs. Calculated Measures

A common question asked by people new to data modelling with DAX (and some not so new) is along the lines of "When should I use a calculated measure and when should I use a calculated column?" This is a fair question and confusion can be caused by the fact that usage of these can overlap. It's possible to show the same value in a report that comes from a calculated column as comes from a calculated measure.

The main difference between calculated columns and calculated measures is the point in time the calculation is executed and how the result of the calculation is stored.

Calculated Columns

With calculated columns, the DAX expression used by the calculation is only executed when data for the table is refreshed. The expression is calculated once for every row in the table, and the single value generated by the calculation is stored as the value in the column for that row and it cannot be changed. It's as if the value generated by the calculated column existed in the source data that was used to import for the table.

Consider calculated columns as part of the data-load process. These calculations execute as one of the very last steps of the data-load process and no user interaction is involved.

Complex calculated columns over larger data sets make data refreshing take longer. If the data model is configured to refresh once a day in the early hours of the morning, however, this should have only a minor impact on users of your reports.

The values generated by calculated columns are stored physically in the data model, which means columns added to very large tables may have a noticeable effect on the memory size of the data model.

Calculated columns are row based, meaning they can quickly perform calculations if all the information required is contained within the same row. Calculations can still use data from rows other than the current row.

Calculated Measures

With calculated measures, the DAX expression could be executed every time a page loads or when a user changes a filter or slicer selection. The DAX expression contained within the calculated measure is executed once for every value that uses it in a report. If a calculated measure is used in a table or matrix visual, it is executed as many times as there are cells in the grid that use it to show a value. When one is used on a line chart visual with 100 points in a series, it executes 100 times for each point, with each execution having a slightly different filter context.

Calculated measures are dynamic and respond to user interaction. They recalculate quickly and often but do not store output in the data model, so they have no impact on the physical size of the data model. Increasing the number of calculated measures in your data model has no impact on speed or size of the model at rest.

Calculated measures are quick to process calculations using a single column over many rows, but you can still write them in a way that allows row-based data processing using iterators.

To better understand the difference between a calculated column and a calculated measure, let's look at how a report requirement might be handled using both options. In Listing 9-8 the requirement is to show a cumulative sum over the 'Daily Summary' [Sum of Revenue] column from the recently created calculated table in Listing 9-7.

Listing 9-8. Cumulative Sum as Calculated Column Is Added to the Calculated Table in Listing 9-7

```
Calculate Revenue as Column =
    CALCULATE(
        SUM('Daily Summary'[Sum of Revenue]),
            FILTER(
            ALL('Daily Summary'),
            'Daily Summary'[Date]<=EARLIER('Daily Summary'[Date])
            )
        )
```

This calculation executes as many times as there are rows in the 'Daily Summary' table. The FILTER function allows each execution of the SUM function to use a different set of values each time. The execution for the row on the oldest day (January 1, 2016) only presents one row to the SUM function. The execution of the SUM expression for the row for January 2, 2012, uses two values. These two executions can take place in any order and still produce the same result.

Once the calculated column has been added, the data model is now larger. In this case, all values in the new column are unique, so there is very little data compression.

Now consider the same calculation but as a calculated measure (Listing 9-9).

Listing 9-9. Cumulative Sum as a Calculated Measure Added to the Calculated Table in Listing 9-7

```
Calculate Revenue as Measure =
    CALCULATE(
        SUM('Daily Summary'[Sum of Revenue]),
        FILTER(
            ALLSELECTED('Daily Summary'[Date]),
            'Daily Summary'[Date]<=MAX('Daily Summary'[Date])
            )
        )
```

In this case, adding the calculated measure to the model doesn't execute the code. The calculated measure remains dormant until it is used in a visual. Once the calculated measure is used in a visual, it is executed once for every value it needs to generate for a visual. Typically this is the number of items in an axis, row, or column header.

The syntax of the calculated measure and column are very similar. Both use the CALCULATE function and the same core SUM expression over the [Sum of Revenue] column. Both apply an instruction to tell the DAX engine what data can be used by the SUM function.

Finally, when both are added to a table visual along with the 'Daily Summary'[Date] field, you see the result in Figure 9-7.

Date	Calculate Revenue as Measure	Calculate Revenue as Column
1 January 2016	1,963.43	1,963.43
2 January 2016	4,942.96	4,942.96
3 January 2016	7,753.67	7,753.67
4 January 2016	10,432.12	10,432.12
5 January 2016	13,449.75	13,449.75
6 January 2016	16,831.63	16,831.63
7 January 2016	20,084.19	20,084.19
8 January 2016	22,172.52	22,172.52
9 January 2016	23,805.00	23,805.00

Figure 9-7. *Output of the code from Listings 9-8 and 9-9*

The two columns produce an identical result for each row in the table. The [Calculate Revenue as Column] simply retrieves data directly from the column in memory and no DAX computations are required. Each cell in the [Calculate Revenue as Measure] column runs the code in the calculated measure. Each execution of the measure has a slightly different query context. In this case the query context is from the row header filtering on each [Date].

If you analyze both approaches using DAX Studio, you start to see how the two differ in terms of overall performance.

The first test uses the calculated column to see how much work was required by the model to produce values for the visual. The option to clear the cache on each run is turned on as is the ability to view the server timings and query plan (Listing 9-10).

Listing 9-10. DAX Query to Test the Performance of Listing 9-8 Using DAX Studio

```
EVALUATE
    SELECTCOLUMNS(
        'Daily Summary'
        , "Date", 'Daily Summary'[Date]
        , "Calculate Revenue as Column", 'Daily Summary'[Calculate
        Revenue as Column]
        ) ORDER BY [Date]
```

Running this calculation ten times over a cold cache yields an average total execution of around 12 milliseconds. As you can see in Figure 9-8, the majority of that output time (usually 90 percent) was spent in the formula engine with a very small amount being spent in the storage engine.

Total	SE CPU
10 ms	0 ms
	x0.0

▨ FE	▧ SE
9 ms	1 ms
90.0%	10.0%

SE Queries	SE Cache
1	0
	0.0%

Figure 9-8. *Server timing output of code from Listing 9-10*

Listing 9-11 shows the same test using the calculated measure instead of the calculated column. Once again, DAX Studio is used to capture the server timings of the query over a cold cache.

Listing 9-11. DAX Query Testing the Performance of Listing 9-9 Using DAX Studio

```
EVALUATE
    SELECTCOLUMNS(
        'Daily Summary'
        , "Date", 'Daily Summary'[Date]
        , "Calculate Revenue as Measure", [Calculate Revenue as Measure]
        ) ORDER BY [Date]
```

This code produces identical output except it uses a calculated measure instead of a calculated column. Running the query many times yielded average query times around the 1.4 second mark (1,400 milliseconds), usually with 99 percent of that being spent in the formula engine (Figure 9-9).

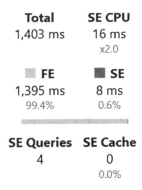

Figure 9-9. *Server timing output of the code from Listing 9-11*

Although this is slower, it is no surprise. The calculated column has the advantage of being precalculated. The time spent by the calculated column to generate the data for the column in the first place is around the same as what we see here with the calculated measure.

Let's look at the two calculations and compare them on the flexibility front. After adding a relationship between the 'Dates' and 'Daily Summary' tables, a slicer is added to the same report page that uses the 'Dates'[Year] column. When this slicer is set to 2017, the result in Figure 9-10 can be observed using two visuals.

Date	Calculate Revenue as Measure	Calculate Revenue as Column
1 January 2017	2,285.68	1,075,075.21
2 January 2017	3,228.67	1,078,303.88
3 January 2017	3,091.53	1,081,395.41
4 January 2017	5,158.53	1,086,553.94
5 January 2017	1,114.94	1,087,668.88
6 January 2017	4,606.07	1,092,274.95
7 January 2017	2,373.62	1,094,648.57
8 January 2017	2,682.22	1,097,330.79
9 January 2017	3,655.13	1,100,985.92
10 January 2017	2,475.18	1,103,461.10
11 January 2017	2,921.63	1,106,382.73
12 January 2017	1,450.61	1,107,833.34
Total	1,082,986.16	

● Calculate Revenue as Column   ● Calculate Revenue as Measure

Year
☐ 2016
■ 2017
☐ 2018

Figure 9-10. *Table and line chart visual showing the results from Listing 9-8 and 9-9 with the slicer set to 2017*

First, the table visual shows that the calculated measure is now showing different values to the calculated column. The reason for this is the values being generated by each new execution of the calculated measure now have a different filter context than they did before. The SUM expression within the calculated measure no longer considers rows that belong to 2016 and 2018. The effect of this additional filtering means the values have been reset and are now only cumulative from the start of 2017.

The column simply outputs the same precalculated values as before. Nothing has changed except the slicer simply stops the visuals from plotting values for dates other than those in 2017.

So, let's revisit the question I posed at the beginning of this section: "When should I use a calculated column vs. a calculated measure?"

If you want the fastest result and have little need to slice and dice data in different ways, a calculated column is a good choice. If you want to have values that react and respond to user interaction, then calculated measures are a good choice.

If you have a problem with the speed and performance of an interactive report, the best place to start looking is probably the code used in your calculated measures rather than the code from your calculated columns.

Show All Sales for the Top Ten Products

To extend on the data you've built so far, let's look at a few examples of how to build some additional summary tables using different techniques. The first example creates a table that holds every sale record but just displays the top ten products by revenue. The first step is to identify which products should be considered top-ten products. Then, using this list, a new table is created that uses all sales filtered for just these products.

The DAX for the calculated table is shown in Listing 9-12.

Listing 9-12. Calculated Table Showing Sales for the Best Ten Products

```
All Sales for Top 10 Products =

VAR InnerGroup =
    SUMMARIZECOLUMNS(
        -- Group BY --
        'Sales'[Product],
        -- Aggregation Column --
        "Sum of Revenue", SUM('Sales'[Total])
        )

VAR Top10PRoducts =
    TOPN(
        10,
        InnerGroup,
        [Sum of Revenue],
        DESC
        )

RETURN
    NATURALINNERJOIN (
        'Sales',
        Top10PRoducts
        )
```

The first variable in this calculation uses the SUMMARIZECOLUMNS function to create a table expression aggregated over the 'Sales'[Product] column. This produces an unsorted working table with one row per product. Each product carries a value in the "Sum of Revenue" column that shows an aggregate of total sales over all time. This might be an opportunity to tweak the filtering so it only considers a date range that is recent and relative to the current period—for example, you might revise it so it only considers sales for the last three months.

The second step uses the TOPN function to identify and return the top ten rows from the InnerGroup table expression when they are sorted by [Sum of Revenue] in descending order. The TOPN function is a filter function, so it returns a table with the same number of columns as in the InnerGroup variable to the Top10Products variable but in this case only ten rows.

You can tweak this function to look for a different number of products at the top or bottom using the ordered table. To find the products with the least sales, simply change the order by parameter from DESC to ASC.

An alternative to using the TOPN function is to assign a ranking column to the InnerGroup variable, which can then use the value in the ranking column to filter the table. You could use the DAX example in Listing 9-13 as an alternative. Figure 9-11 shows the output from this listing.

Listing 9-13. Alternative DAX That Could Be Used in Listing 9-12 in Place of the TOPN Function

```
VAR AddRankColumn =
    ADDCOLUMNS(
        InnerGroup,
        "My Rank",RANKX( InnerGroup, [Sum of Revenue] )
        )

VAR Top10PRoducts =
    FILTER(
        AddRankColumn,
        [My Rank]<=10
        )
```

Product	Sum of Revenue	My Rank ↑
XXX-2	38145.22	1
GGG-4	37506.38	2
AAA-5	36709.12	3
MMM-1	36226.15	4
XXX-5	36100.97	5
HHH-1	33987.91	6

Figure 9-11. *Sample output of the code in Listing 9-12 using the rank column added in Listing 9-13*

An advantage of adding a rank column like the one in Figure 9-11 instead of the TOPN function is that you can use the data in additional calculations such as for performing comparisons of ranking between periods. For instance, you could create two working tables that show an aggregation and ranking of sales for products over contiguous periods. You could then generate a filtered table to show products that have improved (or deteriorated) between the two periods.

This variation uses the FILTER function over the AddRankColumn table expression. If you want to identify the bottom items, the RANKX function takes an optional parameter to sort the data DESC or ASC. This allows the FILTER function to use the <=10 predicate to find the products with the lowest values for sales.

At this point, both versions have a table expression with just ten rows that should represent the products with the best sales.

The last step is to return a copy of the raw sales data filtered for just the products contained in the Top10Products variable. The NATURALINNERJOIN function works well for this with the final RETURN statement performing the equivalent of a T-SQL INNER JOIN between the 'Sales' table and the table expression contained in the Top10Products variable.

The NATURALINNERJOIN function automatically creates join criteria using columns that exist in both tables that have the same name, datatype, and lineage. Multiple column can be used in the join, but in this case, only the [Product] column meets all three requirements. Because the [Product] column in the Top10Products variable can be traced back to the sales table as its origin (via several working variables), it meets the lineage requirement.

The CALCULATETABLE function can be used in place of NATURALINNERJOIN. The parameter signature is the same for both functions. NATURALINNERJOIN is performed in the formula engine, while CALCULATETABLE ('Sales', Top10Products) pushes the filtering operation to the storage engine.

Columns other than [Product] from both tables used in the join are retained. The SELECTCOLUMNS function can be used to control which columns are returned by removing or renaming any unwanted columns.

A sample of the final table should look something like Figure 9-12.

Date	Product	Qty	Price	Total	Sum of Revenue
1/01/2016 12:00:00 AM	XXX-2	1	106.57	106.57	45752.35
1/01/2016 12:00:00 AM	FFF-1	2	464.13	928.26	33037.84
2/01/2016 12:00:00 AM	WWW-5	2	673.69	1347.38	35038.71
3/01/2016 12:00:00 AM	NNN-1	2	796.81	1593.62	34074.43
4/01/2016 12:00:00 AM	CCC-5	1	478.38	478.38	32627.46
4/01/2016 12:00:00 AM	RRR-4	2	708.17	1416.34	33964.24
10/01/2016 12:00:00 AM	NNN-1	1	915.42	915.42	34074.43
10/01/2016 12:00:00 AM	NNN-1	1	673.94	673.94	34074.43
12/01/2016 12:00:00 AM	FFF-1	1	400.55	400.55	33037.84

Figure 9-12. *Sample output of the calculated table in Listing 9-12*

The values in these columns are all based on random numbers so it's the structure that should match your version rather than values.

This calculated table is rebuilt each time your data model is refreshed. Once built, it provides you with a smaller and faster table to use with visuals that only need to work with the bestselling products. This is especially useful if the number of rows in the 'Sales' table grows to be very large.

Double Grouping

A more complex scenario might involve a requirement to perform two or more layers of summarization over data using multiple working tables to generate a new summary table.

Consider a requirement to show a summary table with one row for every product in the sales table, along with the average value for the best ten days for each value. The challenge here is that the best ten days involve different days for each product. Once the best ten days are known for each product, an average can then be generated.

The following technique makes use of these:

- SUMMARIZECOLUMNS

- GENERATE

- GROUPBY and CURRENTGROUP

- MAXX iterator function

- An interesting technique to derive a RANK

The first step of several is to create a working table (Listing 9-14) that will be used as a summary of the daily sales for each product. The raw data may contain multiple transactions for the same product on the same day. This will be the first layer of summarization performed.

Listing 9-14. Summary Calculated Table of the Sales Table in Listing 9-6

```
VAR InnerGroup =
     SUMMARIZECOLUMNS(
                -- Group BY --
                'Sales'[Product],
                'Sales'[Date],
                -- Aggregation Column --
                "Daily Revenue", SUM('Sales'[Total])
                )
```

The InnerGroup variable stores a three-column table showing daily sales for every product for every day it has sales. The next step is to identify which days happen to be the best ten days for each product. This normally involves applying a product-specific ranking value over the [Daily Revenue] column. The RANKX iterator would provide a way to add a column to the table at this point that ranks each row in relation to every other row in the table. But what you need here is a way to have a column that ranks each day of sales *per product*.

One technique is to apply the three steps in Listing 9-15 to the calculation.

Listing 9-15. Additional DAX for the Code in Listing 9-14

```
VAR CopyOfSummaryTable =
     SELECTCOLUMNS(
                InnerGroup,
                "ProductA",[Product],
                "DateA",[Date],
                "Daily RevenueA",[Daily Revenue],
                "RowCounter", 1
                )
```

```
VAR CrossJoinTables =
     FILTER(
                  GENERATE( CopyOfSummaryTable, InnerGroup ),
                  [Product] = [ProductA] &&
              [Daily Revenue]<=[Daily RevenueA]
                  )

VAR ProductByDateRanking =
     GROUPBY(
                  CrossJoinTables,
                  [Product],
                  [Date],
                  "Daily Revenue", MAXX(
                                    CURRENTGROUP(),
                                    [Daily Revenue]
                                    ),
                  "Rank",        SUMX(
                                    CURRENTGROUP(),
                                    [RowCounter]
                                    )
                  )
```

In this code, the CopyOfSummaryTable variable is used to store a copy of the InnerGroup variable. The [Product], [Date], and [Daily Revenue] columns are renamed using SELECTCOLUMNS to avoid a clash over column names when this version of the table is joined back to the InnerGroup table in the next step. The columns are simply renamed to have the "A" character appended.

A new column called [RowCounter] is also added at this step and it is hardcoded to be 1. This column is going to provide the basis for the ranking calculation. Each row is duplicated for every row found for the same product that has a higher value for [Daily Revenue]. Once the duplication has happened, a SUM over the [RowCounter] generates a value that represents the rank.

The next step applies a cartesian join between the InnerGroup and the CrossJoinTables table variables. A FILTER function makes sure that only rows from InnerGroup are matched with rows from CopyOfSummaryTable that have the same value for [Product]. The [Daily Revenue] <= [Daily RevenueA] logic allows for as many rows from the CopyOfSummaryTable variable that have the same or higher value for [Daily Revenue] from the InnerGroup variable.

For any given product in the InnerGroup table variable, the row that happens to have the highest value for [Daily Revenue] for that product only finds one matching row from CopyOfSummaryTable (it finds itself). The InnerGroup row for the same product with the next highest value finds two matching rows, and so on, until the row from InnerGroup with the lowest value for the same product is the only row left unmatched. This row is then matched with every other row from CopyOfSummaryTable for the same product, meaning a SUM over the [RowCounter] column returns a number that has the highest number (lowest ranking).

The last step in this section is to group the cartesian output of the previous step back to a single row per product and day. Neither the SUMMARIZE nor the SUMMARIZECOLUMNS functions can be used over a table expression stored in a variable. Both functions only work with a physical table. Fortunately, the GROUPBY function can be used, but aggregation expressions must use iterator functions. This creates a summary table over the CrossJoinTables variable grouping by [Product] and [Date].

There is no COUNTROWSX iterator function in DAX; if it existed, it could be used to count the number of rows belonging to each [Product] and [Date] while they were being summarized to provide a value for rank. In this case, the [RowCounter] column introduced in the first step allows an equivalent approach to be taken by performing a SUMX over the hardcoded values in this column.

If no BLANK values are in the column, an alternative is to simply COUNTAX(CURRENTGROUP(), [Column]) to avoid introducing the column that has a value of 1 in every row.

The CURRENTGROUP() function used in both the MAXX and SUMX functions is a reference back to the table used as the first parameter in the GROUPBY function.

Here the MAXX function is being used to allow the daily sales total for each day to be stored in the [Daily Revenue] column. This helps reduce duplicated instances of the same value back to a single instance for each product/day. This value does not reflect the max of individual transactions, rather it represents the combined sales total for the product across each day. This is not the step that performs the aggregation; instead it helps preserve the daily sales data through to the next step.

The output of the ProductByDateRanking variable would look something like Figure 9-13.

Product	Date	Daily Revenue	Rank ↑
AAA-2	27/04/2016 12:00:00 AM	2314.38	1
AAA-2	6/03/2017 12:00:00 AM	2226.57	2
AAA-2	5/09/2017 12:00:00 AM	2214.69	3
AAA-2	22/12/2016 12:00:00 AM	1785.36	4
AAA-2	20/05/2018 12:00:00 AM	1182.28	5
AAA-2	27/06/2017 12:00:00 AM	1017.99	6
AAA-2	5/07/2017 12:00:00 AM	976.23	7
AAA-2	4/07/2018 12:00:00 AM	959.91	8

Figure 9-13. *Sample output of the cod in Listings 9-14 and 9-15 ordered by rank*

When sorted by product and rank, the daily revenue data appears as expected. Notice the dates for each row jump around. Each product has different days that make up the different ranking positions.

This table carries a ranking value for any day in which the product registers a sale. The next step is to filter this data down to be just the top ten rows by rank for each product. Listing 9-16 shows this straightforward use of the FILTER function.

Listing 9-16. Additional Step to the Code in Listing 9-15

```
VAR TopTenDaysPerPRoduct =
     FILTER(
          ProductByDateRanking,
          [Rank]<=10
          )
```

Now you have a set of data that only covers the ten best days by product. The last step to meet the initial requirements is to create an average over the [Daily Sales] per product. Again, remember that SUMMARIZE and SUMMARIZECOLUMNS cannot be used here because they do not work with table expressions stored in variables. So, the last step (Listing 9-17) uses the GROUPBY function to perform the average.

Listing 9-17. Final RETURN Statement Added to the Code in Listing 9-16

```
RETURN OuterGroup =
            GROUPBY(
                 TopTenDaysPerProduct,
                 -- Group BY --
                 [Product],
                 -- Aggregation Column --
                 "Average of Top 10 Best Days",
                 AVERAGEX(
                                 CURRENTGROUP(), [Daily Revenue]
                                 )
                 )
```

The step now aggregates the working data from the previous step using the TopTenDaysPerProduct variable down to just one row per product. All products will have a row in this table. Then the AVERAGEX iterator function is used in conjunction with the CURRENTGROUP() function to perform an average over the [Daily Revenue] data for each product.

The final output is a two-column table (Figure 9-14) that meets the unusual reporting requirement. I didn't choose this scenario because it is likely to be a frequent problem you encounter. Instead, I chose it because it showcases a variety of techniques that you can use in DAX to address scenarios that have an extra degree of complexity.

Product	Average of Top 10 Best Days
AAA-1	1337.812
AAA-2	1667.731
AAA-3	994.838
AAA-4	1114.615
AAA-5	1252.865
BBB-1	1324.581
BBB-2	1356.458

Figure 9-14. *Sample of accumated code in Listings 9-14–9-17*

Index

259

© Philip Seamark 2018
P. Seamark, *Beginning DAX with Power BI*, https://doi.org/10.1007/978-1-4842-3477-8

Get the eBook for only $5!

Why limit yourself?

With most of our titles available in both PDF and ePUB format, you can access your content wherever and however you wish—on your PC, phone, tablet, or reader.

Since you've purchased this print book, we are happy to offer you the eBook for just $5.

To learn more, go to http://www.apress.com/companion or contact support@apress.com.

Apress®

All Apress eBooks are subject to copyright. All rights are reserved by the Publisher, whether the whole or part of the material is concerned, specifically the rights of translation, reprinting, reuse of illustrations, recitation, broadcasting, reproduction on microfilms or in any other physical way, and transmission or information storage and retrieval, electronic adaptation, computer software, or by similar or dissimilar methodology now known or hereafter developed. Exempted from this legal reservation are brief excerpts in connection with reviews or scholarly analysis or material supplied specifically for the purpose of being entered and executed on a computer system, for exclusive use by the purchaser of the work. Duplication of this publication or parts thereof is permitted only under the provisions of the Copyright Law of the Publisher's location, in its current version, and permission for use must always be obtained from Springer. Permissions for use may be obtained through RightsLink at the Copyright Clearance Center. Violations are liable to prosecution under the respective Copyright Law.

Printed in the United States
By Bookmasters